Emotional Wellbeing in the Domestic Dog

Dedication

For Clive Pedley, the man who taught me to love dogs.
For all my clients, thank you for choosing me.
And for Cheddar, my heart, the dog that changed my life.

Emotional Wellbeing in the Domestic Dog

Jade Nicholas
Clinical Animal Behaviourist (APBC/ABTC)

CABI is a trading name of CAB International

CABI
Nosworthy Way
Wallingford
Oxfordshire OX10 8DE
UK

CABI
200 Portland Street
Boston
MA 02114
USA

Tel: +44 (0)1491 832111
E-mail: info@cabi.org
Website: www.cabi.org

T: +1 (617)682-9015
E-mail: cabi-nao@cabi.org

© Jade Nicholas 2025. All rights, including for text and data mining, AI training, and similar technologies, are reserved. No part of this publication may be reproduced in any form or by any means, electronically, mechanically, by photocopying, recording or otherwise, without the prior permission of the copyright owners.

The views expressed in this publication are those of the author(s) and do not necessarily represent those of, and should not be attributed to, CAB International (CABI). Any images, figures and tables not otherwise attributed are the authors(s)' own. References to internet websites (URLs) were accurate at the time of writing.
CAB International and, where different, the copyright owner shall not be liable for technical or other errors or omissions contained herein. The information is supplied without obligation and on the understanding that any person who acts upon it, or otherwise changes their position in reliance thereon, does so entirely at their own risk. Information supplied is neither intended nor implied to be a substitute for professional advice. The reader/user accepts all risks and responsibility for losses, damages, costs and other consequences resulting directly or indirectly from using this information.

CABI's Terms and Conditions, including its full disclaimer, may be found at https://www.cabi.org/terms-and-conditions/.

A catalogue record for this book is available from the British Library, London, UK.

ISBN-13: 9781836992431 (hardback)
9781836990703 (paperback)
9781836990710 (ePDF)
9781836990727 (ePub)

DOI: 10.1079/9781836990727.0000

Commissioning Editor: Alexandra Lainsbury
Editorial Assistant: Emma McCann
Production Editor: Rosie Hayden

Typeset by Exeter Premedia Services Pvt Ltd, Chennai, India
Printed in the USA

Contents

Contributors		xi
Acknowledgements		xiii
Preface		xv
1	**Emotional Wellbeing and Dogs**	**1**
	Jade Nicholas	
	Introduction	1
	The UK Dog Demographic	2
	Expectation Versus Reality	3
	Cost	3
	Behaviour	4
	Conclusion	4
	References	5
2	**Emotional Wellbeing and Animal Ethics**	**6**
	Sophie East and Jade Nicholas	
	Introduction	6
	What is Welfare (And How Do We Measure it)?	7
	The Five Welfare Needs	8
	The Animal Welfare Assessment Grid (AWAG)	8
	Maslow's Hierarchy of Needs	9
	Animal Ethics	11
	Conclusion	11
	References	12
3	**The History of Dogs**	**13**
	Sophie East and Jade Nicholas	
	Introduction	13
	The Domestication of Dogs	14
	Changes in Genetics	14
	Glucocorticoids	15
	Oxytocin	15
	How have these genetic changes affected domestication?	15
	Behavioural Changes Through Domestication	16
	Modern Implications of Domestication	17
	Free-Ranging Dogs	18
	Conclusion	19
	References	19

4	**Emotional Wellbeing Starts with the Right Dog**	21
	Jade Nicholas	
	Introduction	21
	Choosing a Dog Breeder	22
	Rescue and Rehoming	23
	The 'Right' Breed	23
	Physical considerations	23
	Behavioural implications	24
	Natural behaviours	25
	Conclusion	26
	References	27
5	**Traumatized Dogs**	28
	Jade Nicholas	
	Trauma in Dogs	28
	General Trauma-Informed Care Principles	29
	Safety	29
	Connections	30
	Managing emotions	30
	Overseas Rescue Dogs	31
	Behaviour and Welfare Concerns in Dogs Imported From Overseas	31
	Puppy Farms	33
	Dogs Impacted by Criminal Activity	36
	Supporting a Traumatized Dog (Using the 'Three Pillars of Trauma-Informed Care')	36
	Pillar 1: Safety	36
	Pillar 2: Connections	39
	Pillar 3: Managing emotions	40
	Case vignette: Jerry	41
	Real-Life Application	42
	References	42
6	**Emotional Wellbeing While Learning**	44
	Jade Nicholas	
	Introduction	44
	How Dogs Learn	45
	Classical conditioning	45
	Operant conditioning	47
	Commonly Used Training Methods	48
	Positive reinforcement versus positive punishment	49
	LIMA and LIFE Methods	50
	Increasing meaningful choices	50
	Identify behavioural functions	52
	Maximize training success	52
	LIMA versus LIFE: conclusions	52
	Equipment	52
	Ethical considerations	53

Physical considerations 54
Electronic collars (e-collars) 54
An Optimum State for Learning 55
Arousal and emotional valence 55
Trigger stacking 57
Distractions 57
Conclusion 58
References 58

7 **Emotional Wellbeing in the Home** **61**
Jade Nicholas
Introduction 61
Relationships with Humans 61
Babies and children 63
Visitors 66
Routine and Predictability 68
Environmental unpredictability 68
Changes for the better 68
Case vignette: Twiglet 69
Leaving a Dog Alone 69
Relationships with Other Animals 70
Dogs living with other dogs 70
Dogs living with cats 73
Dogs living with other animals 74
Indoor Environment/Safe Space 74
Enrichment 74
References 75

8 **Emotional Wellbeing Outdoors** 77
Jade Nicholas
Introduction 77
Exhibiting Natural Behaviours 77
Sniffing 78
Canine predatory sequence 78
Breed-specific behaviours 79
Fun for Everyone 81
The Dangers of 'Fetch' 81
So, to fetch, or not to fetch? 82
Agoraphobia and Walk Refusal 83
Building Confidence (Inspired by Free Work) 83
Muzzle Training 84
Choosing the right muzzle 85
Introducing the muzzle 85
Appropriate Greetings 88
Unfamiliar dogs within a home 89
Unfamiliar dogs outdoors 89
People 91

A Note on 'Doggy Day-Care' and Dog Walkers	92
Doggy day-care	92
Dog walkers	93
Coping with Reactivity	94
What is 'reactivity'?	94
Adjusting your mindset	95
Exercises for engagement	95
Desensitization and counter-conditioning	97
To walk, or not to walk?	98
A message to caregivers	99
Car Travel	99
Reducing travel	100
Safety signals	100
Comfortable access	101
Games for the car	101
Desensitization	102
Additional measures	102
Conclusion	103
References	103

9   The Link Between Physical and Emotional Wellbeing   106
Jade Nicholas and Fabian Rivers

Introduction	106
'A Vet's Perspective' by Fabian Rivers	107
Pain, discomfort and sadness. Empathy from a vet's perspective	107
The evolving landscape of veterinary practice	107
Obesity	107
Osteoarthritis: the unseen burden of movement	108
Dermatological pain: the itch that speaks volumes	108
Breeding, conformation and surgical birth	109
Behavioural change as a red flag for pain	110
Economic pressures and the veterinary dilemma	110
Reflections and the road ahead	111
Pain and Wellbeing	111
Pain and Behaviour	112
Identifying pain in a behaviour case	113
Behaviour and musculoskeletal pain	113
Case vignette: Mara	114
Behaviour and gastrointestinal issues	115
Behaviour and cognitive health	115
Improving Welfare During Handling	116
Handling in the veterinary clinic	116
Low Stress Handling®	116
Optimizing the environment	117
Supporting dogs through consultation	117
Handling at the grooming salon	118
Guidance for groomers	119

Desensitization and Co-operative Care	119
What is co-operative care?	120
Limitations	120
Teaching a co-operative care protocol	120
Suggested desensitization exercises	122
Conclusion	124
References	124
Appendix A	127
Index	**129**

Contributors

Jade Nicholas is an APBC Clinical Animal Behaviourist. She holds a BA (Hons) in Animal Welfare and Society, an MSc in Clinical Animal Behaviour, and previously worked for Dogs Trust on the Behaviour Support Line. She now runs her own company, About Your Dog Ltd.

Sophie East is a graduate in Animal Welfare and Society from the University of Winchester. At the time of publishing Sophie works as a Virtual Assistant for About Your Dog Ltd, working towards a career in companion animal care in the near future.

Fabian Rivers MVDR, MRCVS GPCert(ExAP) is a companion animal and exotics vet from Birmingham, formerly winning Young Vet of the Year in 2020. His most popular work has been shown on the BBC and Channel 4, documenting animal welfare (including the breeding and health of various species), their behaviour, and the associated risks of these issues to the wider community.

Acknowledgements

Thank you to Jericho Reid and Charlie Nicholas for the beautiful photographs as well as the clients who kindly allowed me to photograph their dogs.

Thank you to every science-based professional that has taught me, supported me and helped me to grow through my years of study and practice. Special thanks are in order to the team (past and present) at the Professional Association of Canine Trainers (PACT) and the Association of Pet Behaviour Counsellors (APBC) as well as the staff at the University of Lincoln.

Preface

This book was inspired by a visit to a coffee shop. This particular coffee shop is set in the beautiful countryside of rural Hampshire, nestled into a barn on the outskirts of a farm. For dogs, this coffee shop offers endless possibilities for sniffing, socializing and relaxing in the fresh air and sunshine. However, on this particular visit, I could not concentrate on enjoying my coffee and cinnamon bun, for I was distracted by a vizsla and their human.

This vizsla had been brought along to this beautiful coffee shop with his humans but appeared to be nothing more than an inconvenience to them. Although he was already wearing a ghastly figure of eight slip lead secured up on the sensitive part of his neck, he was restricted further by being kept at heel for the *entire duration of their visit*.

The dog was yanked back to his seated position every time he went to sniff, wag his tail at another dog or (heaven forbid) tried to enjoy himself. He wasn't doing anything wrong or dangerous, he was just being a dog. I considered whether in a few months or years' time an enquiry would land in my inbox to request help for this dog's behaviour. Perhaps his humans would come to tell me that after years of dragging him into line and restricting his autonomy he had delivered a serious bite, and that if I couldn't fix it he would be looking for a new home (or worse).

Knowing that unsolicited advice is rarely received happily, I did not intervene. In fact, if I intervened every time I saw a violation of dog's emotional wellbeing I would never get anything done. Instead, I decided to put my thoughts onto paper in the hope that people will educate themselves.

This book is written for every dog that is expected to 'know their place' and blindly submit to humans. It is my wish to rlease as many dogs and caregivers as possible from the grip of 'obedience' and to help nurture a meaningful, secure relationship that improves quality of life for both. Dogs are sentient beings with an individual lived experience; we must treat them as such.

Emotional Wellbeing and Dogs

Jade Nicholas*

CAB, Winchester, UK

Abstract

This chapter sets the tone for the rest of the book, introducing the dog and caregiver demographic in the UK as well as highlighting some of the problems facing them. In particular, the expectations of dog ownership versus the reality are explored alongside the strict regulations we place upon our dogs as a result.

Introduction

> We walk together, friends, enjoying each other's company. Taking our time to watch the birds, to smell the scent of wafting muffins in the café, to have small passing visual conversations with other dogs. Our outings are for her, for us. They are not for me alone, not for the child who wants to place his hands on her to stroke her soft fur, not for the people who smile and tell me how pretty she is, and not for the elderly woman who stops to ask about how old she is or what her name is. This time is for her to exist without anyone invading her space, without the threat of unsolicited touching. I will never ask her to disengage with smells or not look at the dog across the road. I will never shorten the lead and force her to move in a continuously straight line at a continuously human pace. It doesn't mean there are no guidelines, and it doesn't mean she has entirely free will to do as she pleases. We are sharing our space with others who also matter, after all. That's why I have provided her with skills to feel safe and be successful around others. And so we enjoy our time and wind our way back along the river and to our home. I sit down and begin to write, and she burrows into a blanket on the couch, head resting on my pillow, and we both feel contented.
>
> (Jones, 2024)

Jones (2024) so beautifully describes the perfect dog walk in her book *Constructing Canine Consent*. In this complex human world with so many moving parts, it is easy for domestic dogs to become overwhelmed and fall into bad habits, perhaps going on to develop long-lasting behaviour problems. And so, we tighten the leash and restrict them further, expecting them to fall in line and conform to the life we (humans) have forced upon them. But, what if we were to slow down and consider their emotional experience? What if we saw undesirable behaviour as a consequence of emotional dysregulation, and sought to understand how we can make our dogs feel more comfortable? Perhaps then we would no longer require such militant handling or rigorous restrictions.

Too often, frightened or frustrated dogs are placed into uncomfortable situations with the aid of aversive equipment such as prong collars or slip leads, the latter typically wound tightly around their nose. So much is expected of them, and so when they try to communicate that they can no longer cope they are punished even further. Often, those that struggle the most are in chronic undiagnosed pain that would cripple a human, yet we expect them to carry on without making a fuss.

---

*Corresponding author: aboutyourdog@outlook.com

We take traumatized dogs to our local pub or café and chastise them when they bark at a stranger coming too close, forgetting that the only experience they have had with people is being snatched off of the street in their origin country and stuffed into the back of a lorry. We try to fit a quick 20-minute walk into our working day and berate our dog for stopping to sniff too many times, forgetting that they have waited all morning for their favourite activity of the day. They spend years without ever coming into contact with a child, and yet we expect them to integrate perfectly when we bring a screaming newborn baby home with us.

We even discriminate against them based on breed, size or appearance. The frightened Dachshund simply gets an eye roll and perhaps even a giggle when they bark at another dog, and yet it is unacceptable for a Cane Corso to do the same behaviour, for exactly the same reason, and we put a shock collar on him as a result.

Hopefully, the empathetic reader can understand why these expectations and skewed belief systems lead to problems, and why it is so important to consider **emotional wellbeing** in our dogs. From a human perspective, we actually make life harder for ourselves by ignoring such a fundamental piece of the puzzle.

Emotional Wellbeing in the Domestic Dog seeks to unpack the idea that obedience and servitude are key determinants of the dog–caregiver relationship and instead puts emotional wellbeing at the centre of everything. Common scenarios are looked at through a behaviour lens and practical, welfare-friendly solutions are given for both caregivers and professional behaviour practitioners.

The UK Dog Demographic

As this book is written in 2025, there are an estimated 10.6 million dogs in the UK with over one quarter of adults sharing their home with a canine companion (PDSA, 2024). This number is increasing on a yearly basis.

The 2024 Dogs Trust National Dog Survey (Dogs Trust, 2024) revealed that Labrador Retrievers were the number one most popular dog breed (30,000 accounted for in this survey, suggesting an estimated 1 million in the UK) followed by Cocker Spaniels and Border Collies.

In the People's Dispensary for Sick Animals (PDSA) 2023 report, 23% of all pet owners reported that they had not previously had a pet, an increase on previous years, and 42% of those owners had not owned dogs in their adult life (PDSA, 2023). The 2023 report indicated that most people (55%) have 'pedigree' dogs while 28% of caregivers have a crossbreed. The major increase seen in recent years is a rise in designer crossbreeds such as those crossed with poodles. In 2022, the PDSA found that 53% of dogs were found online, and this number has increased, rising to 65% in 2023. It is widely agreed that this is an area of concern, and this will be further investigated in later chapters.

Sadly, there is a large number of animals entering the rehoming system. In 2023, Dogs Trust reported 44,883 relinquishment enquiries with upwards of 10,000 dogs entering their rehoming system alone (Dogs Trust, 2023).

Evidently, the companion animal demographic is continuing to evolve as well as the demographic of the guardians who take them into their homes. So, what is the problem?

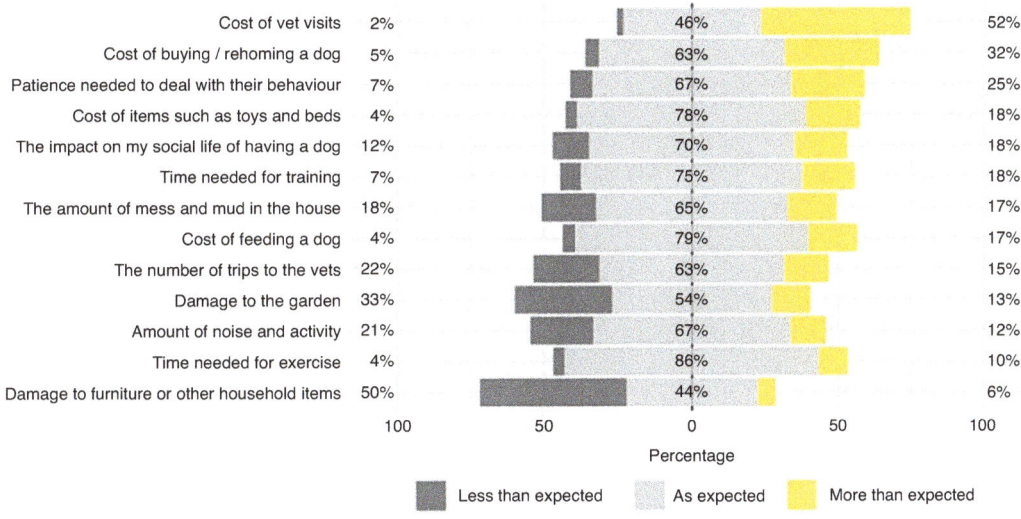

Fig. 1.1. Percentage of respondents (*n* = 3,54,046) stating that their expectations of dog ownership (based on a series of 13 statements) were either less than, more than, or as expected (Anderson *et al.*, 2024). Image is open access.

Expectation Versus Reality

Perhaps the continued rehoming and abandonment crisis can be boiled down to expectations of dog ownership not meeting the reality (Fig. 1.1). Unreasonable expectations are often caused by a lack of understanding or comprehension, and this is very evident in the way that the public cares for their dogs. All too often it is said that dogs should naturally be 'obedient' and 'know who is boss', forgetting that dogs are sentient beings who were not born solely for the purpose of pleasing humans. In a disposable society, dogs that do not meet expectations are for some people simply a burden.

An individual that expects their dog ownership journey to be straightforward is likely due to experience a sobering reality very quickly, possibly becoming frustrated and causing them to partake in some of the counter-productive behaviours and practices described in this book.

Cost

Dogs are expensive. It is expensive to feed them, provide them with veterinary care and to have them trained. Unfortunately, this fact continues to be wildly underestimated, with almost three quarters of pet caregivers failing to look into the cost of their decision prior to getting their dog (PDSA, 2024). In the PDSA report 1 year earlier, 60% of caregivers reported that they had underestimated the monthly cost of owning a pet (PDSA, 2023), and this is thought to be a key contributor to the high level of relinquishment in the UK. One study found that only 30% of respondents (*n*=1814) correctly estimated the realistic lifetime cost of having a dog (Philpotts *et al.*, 2024).

A common barrier for responsible pet ownership is escalating veterinary fees. Concerns have been raised about the cost of veterinary fees, with multiple investigations launched in 2024 (BBC News, 2024) and many pet owners describing vets as 'greedy' and 'money grabbers' (Powell *et al.*, 2022; Anderson *et al.*, 2024). As of 2025, the investigations are still ongoing, and a capping of veterinary fees has been suggested to prevent these issues from continuing. But will it be enough to level expectations?

Given that freedom from sickness and discomfort is a basic welfare requirement for keeping dogs, veterinary fees cannot be a barrier for responsible dog ownership. Those more aware of the potential costs might choose to pay for insurance so as not to unexpectedly fall into financial difficulty or worse, fail to get their dog the treatment it needs.

Behaviour

With around 79.4% of canine caregivers choosing to welcome a dog into their home for companionship (Holland *et al.*, 2022), there is unsurprisingly a disconnect when training and behaviour problems start to arise.

For some, problems can start early with unrealistic expectations of normal puppy behaviour such as toileting inside, chewing and puppy biting. For others, adolescence raises concerns as their bouncy dog becomes bigger and changes into a curious teenager (often with selective hearing). Sometimes, the behaviour problems occur relatively spontaneously, or for others they have always been there, such as with dogs adopted later in life. Inability to cope with these behaviours is a driving force behind many cases of relinquishment (Bouma *et al.*, 2020) as well as being the motivation for many people to turn to aversive, punishment-based training methods. Many caregivers choose to adopt from shelters, but it is not uncommon for dogs to end up returned due to behaviour problems (Powell *et al.*, 2022).

It has been suggested in recent research that caregivers find themselves requiring more patience, and more time to dedicate to training than expected, with some indicating that they did not expect training to be a lifelong process (Anderson *et al.*, 2024). Some studies have shown that those who underestimated the work involved with caring for their dog were more likely to report problem behaviours such as barking, chewing and aggressive behaviour (Patronek *et al.*, 1996), and indeed more recent studies have shed light on the fact that many owners did not expect behaviour problems to occur (Anderson *et al.*, 2024). It is also suggested that caregivers with a high expectation of the human-animal bond, perhaps those seeking emotional support, were more likely to return their dogs (Powell *et al.*, 2022).

Interestingly, experienced dog owners are more likely to hold high expectations (Bouma *et al.*, 2020; Anderson *et al.*, 2024), and indeed, people have a tendency to compare their current dog to previous good experiences, forgetting once again that all dogs are individuals with unique worldviews.

Conclusion

> Thus far, dogs have done their best to adjust to the many changes and restrictions we have imposed upon them, in particular our expectation that they will be faithful companions when we need them to be, and unobtrusive when we do not. However, the cracks inherent in this compromise are beginning to widen.
>
> (Bradshaw, 2011)

The UK dog demographic is in crisis. A combination of dogs becoming too easily accessible, and caregivers without the knowledge and skills to care for them properly has led to a rehoming system that is on its knees without a light at the end of the tunnel.

Can this be boiled down to a lack of understanding of canine emotional wellbeing? Is the lived experience of man's best friend something that the majority of people cannot comprehend?

In reading this book, the eyes of both caregivers and practitioners should be opened to the benefits of enhancing quality of life for our dogs, ultimately choosing to provide a life well lived over a life of service to humans.

References

Anderson, K.L., Holland, K.E., Casey, R.A., Cooper, B. and Christley, R.M. (2024) Owner expectations and surprises of dog ownership experiences in the United Kingdom. *Frontiers in Veterinary Science* 11. DOI: 10.3389/fvets.2024.1331793.

BBC News (2024) Vet prescription fees could be capped by watchdog. *BBC News*, 2024. Available at: https://www.bbc.co.uk/news/articles/cx00dl5gwp8o (accessed 5 June 2025).

Bouma, E.M.C., Vink, L.M. and Dijkstra, A. (2020) Expectations versus reality: Long-term research on the dog–owner relationship. *Animals* 10(5), 772. DOI: 10.3390/ani10050772.

Bradshaw, J. (2011) *In Defence of Dogs*. Penguin Books, London.

Dogs Trust (2023) Dogs Trust Annual Report 2023. Available at: https://dt-prod-pdfs1.s3.eu-west-1.amazonaws.com/annual-report-2023-updated.pdf (accessed 24 July 2025).

Dogs Trust (2024) Welcome to the results of the National Dog Survey 2024. Available at: https://www.dogstrust.org.uk/downloads/Dogs_Trust_NDS_Report_2024__.pdf (accessed 10 July 2025).

Holland, K.E., Mead, R., Casey, R.A., Upjohn, M.M. and Christley, R.M. (2022) Why do people want dogs? A mixed-methods study of motivations for dog acquisition in the United Kingdom. *Frontiers in Veterinary Science* 9, 877950. DOI: 10.3389/fvets.2022.877950.

Jones, E. (2024) *Constructing Canine Consent*. CRC Press, Boca Raton, Florida.

Patronek, G.J., Glickman, L.T., Beck, A.M., McCabe, G.P. and Ecker, C. (1996) Risk factors for relinquishment of dogs to an animal shelter. *Journal of the American Veterinary Medical Association* 209(3), 572–581.

PDSA (2023) PDSA Animal Wellbeing (PAW) Report. Available at: https://www.pdsa.org.uk/what-we-do/pdsa-animal-wellbeing-report/paw-report-2023 (accessed 10 July 2025).

PDSA (2024) PDSA Animal Wellbeing (PAW) Report. Available at: https://www.pdsa.org.uk/what-we-do/pdsa-animal-wellbeing-report/paw-report-2024 (accessed 10 July 2025).

Philpotts, I., Blackwell, E.J., Dillon, J., Tipton, E. and Rooney, N.J. (2024) What do we know about dog owners? Exploring associations between pre-purchase behaviours, knowledge and understanding, ownership practices, and dog welfare. *Animals* 14(3), 396. DOI: 10.3390/ani14030396.

Powell, L., Lee, B., Reinhard, C.L., Morris, M., Satriale, D, *et al*. (2022) Returning a shelter dog: The role of owner expectations and dog behavior. *Animals* 12(9), 1053. DOI: 10.3390/ani12091053.

2 Emotional Wellbeing and Animal Ethics

SOPHIE EAST[1] AND JADE NICHOLAS[2]*

[1]Brighton, UK; [2]CAB, Winchester, UK

Abstract

Wellbeing can be achieved by enhancing welfare. However, with ambiguity surrounding both concepts (regarding definition and attainment) it can be difficult to both quantify and measure them. This chapter considers various UK legislations as well as information provided by individual organizations to provide measures and definitions aligning with an animal rights ethical perspective. The specific needs of dogs are discussed using a recent model based on Maslow's Hierarchy of Needs.

Introduction

There is no universally recognized definition of 'emotional wellbeing' in animals, though the area continues to generate interest. Similarly, there is no agreed method of assessment or measurement. However, given similarities between human and canine brain structure as well as emotional states we can already measure, we can consider measures of emotional wellbeing in humans as a loose framework upon which to build. We might start to consider emotional wellbeing in animals as a term to describe their fulfilment or joy, or to consider their sense of safety.

The UK Office of National Statistics (ONS) runs a continuous 'Measuring National Wellbeing Programme' for humans, established in 2010, with the most up-to-date version released in 2019 (Office for National Statistics, 2019). In order to measure wellbeing in people, they consider an individual's feeling regarding the following:

1. Personal wellbeing (opinion based).
2. Our relationships.
3. Health.
4. What we do (satisfaction in work and/or leisure).
5. Where we live.
6. Personal finance.
7. Education and skills.
8. Economy.
9. Governance.
10. Environment (natural environment).

While many of these areas of enquiry are not applicable to dogs, a significant number of them are, and so we can take inspiration from this pre-established system of measurement.

*Corresponding author: aboutyourdog@outlook.com

Dogs value strong relationships, optimum health, a stable routine and environment, and time spent outdoors as much as humans. These specific measures contribute to the experience of pleasure and positive emotions, an important consideration when assessing wellbeing.

It can be tempting to assume 'all is well' if there is no obvious experience of negative emotional states; however, this cannot be the standard we strive to achieve. Instead, we have a responsibility to also ensure that the individual is having an abundance of positive experiences. It is important to both optimize and preserve positive experiences and to reduce negative ones in the life of the domestic dog.

What is Welfare (And How Do We Measure it)?

Measuring the welfare of an animal is a complex task. This is because there is no one definition of welfare or set method for assessing it. The most famous definition of welfare was put forward by Professor Donald Broom: 'the welfare of an individual is its state as regards its attempts to cope with its environment' (Broom, 1998). However, Ohl and Putman defined welfare in a more nuanced way, pertaining to the experience of the individual: 'an animal's [welfare] status must be perceived and judged by that animal itself' (Ohl and Putman, 2018). This more recent suggestion falls in line with the modern school of animal behaviour that welfare is a subjective experience and not for humans to decide. For example, if an animal demonstrates that they are experiencing pain, but we are unable to find the source of that pain, it is not to say that pain is not there.

With the words of Ohl and Putman in mind, how can welfare be assessed? Efforts to measure and improve animal welfare have typically focused on three broad objectives. These objectives are (Fraser, 2009):

1. To ensure good physical health and functioning of animals.
2. To minimize unpleasant affective states such as pain or fear and to allow normal pleasures.
3. To allow animals to develop and live in ways that are natural for the species.

Having objectives such as these help us to reach a desired goal. Without objectives, measuring welfare would be nearly impossible.

To achieve these objectives, it is important to know what to look for. When measuring animal welfare, three primary states are to be considered: the animal's behavioural, physiological and neurological states. By focusing on these three states, we are provided with both bad and good welfare indicators. The most common indicators of welfare are (Broom, 1998):

Poor indicators of welfare:

1. Body damage.
2. Illness.
3. Abnormal behaviours (in the context of that individual).
4. Reduction in fitness.
5. Suppression of normal behaviour patterns.

Good welfare indicators:

1. Variety of normal behaviours expressed.
2. Expression of preference behaviours.

3. Physiological indicators of pleasure.
4. Behavioural indicators of pleasure.

With a definition and objectives on board, how specifically is welfare measured? To this day, there is no generally accepted method for measuring animal welfare due to scientific and philosophical conflicts within the field, and the vastly different welfare needs of each species.

The Five Welfare Needs

The Animal Welfare Act (2006) provides minimum expectations in the form of the 'Five Welfare Needs' (Section 9).
    In the UK, it is a legal requirement to meet the following:

1. Need for suitable environment.
2. Need for a suitable diet.
3. Need to be able to exhibit normal behaviour patterns.
4. Need to be housed with, or apart from other animals.
5. Need to be protected from pain, suffering, injury and disease.

Though these requirements may seem simple to a doting caregiver, the 2024 PDSA PAW Report found that only 17% of owners surveyed were aware of the guidelines (PDSA, 2024). This statistic may not be particularly surprising, since violations of the Animal Welfare Act happen every day in the homes of the nation's dogs. Common violations might include:

- Not providing a safe and secure place to sleep.
- Feeding inappropriately or irregularly.
- Being punished for trying to exhibit natural behaviours such as sniffing, digging or playing.
- Being housed with other animals that make them feel unsafe.
- Ignoring pain or medical concerns, failing to have animals groomed or failing to have them vaccinated against preventable diseases.

Once the welfare needs are met, an animal is meeting minimum requirements as specified by law in the UK. However, much more is required thereafter.

The Animal Welfare Assessment Grid (AWAG)

While the Animal Welfare Act (2006) provides a basic framework by which to assess and maintain welfare, a more advanced measurement tool for assessing canine welfare is the Animal Welfare Assessment Grid (AWAG).

> The AWAG is unique in that is considers the lifetime experience of the individual animal, and the 'cumulative suffering' that can have an impact on quality of life.
>
> (AWAG, 2025)

Though it may not always be necessary to do so, the AWAG is available for use by anyone that provides care to an animal (though at present not all animals are covered), meaning

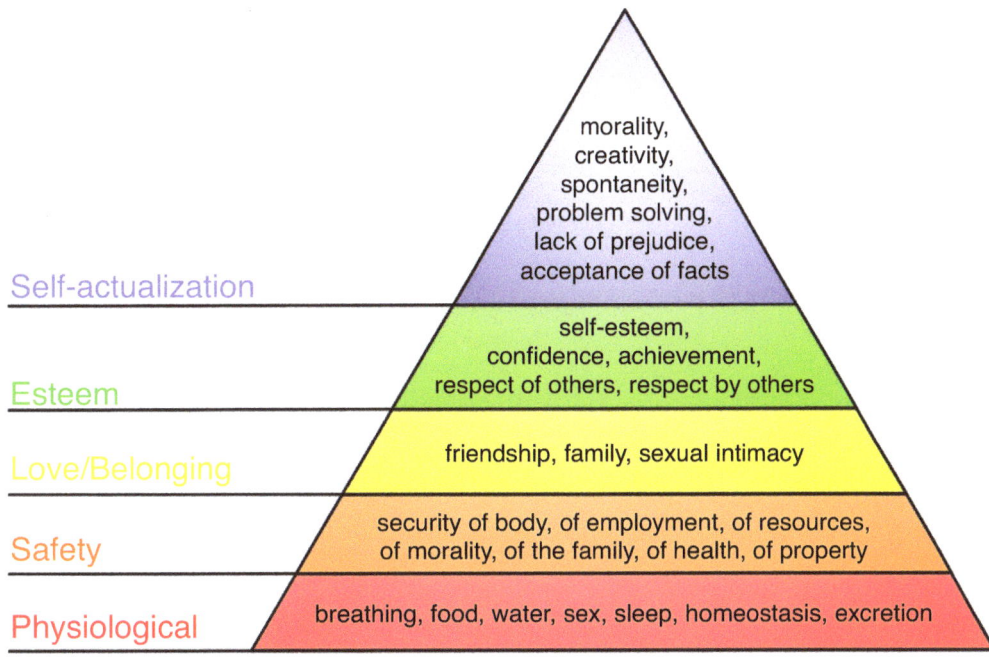

Fig. 2.1. Maslow's Hierarchy of Needs. Image is open access.

that any professional in the canine care sector may benefit from assessing animals this way. Using the AWAG framework benefits animals because it does not overlook the impact past life experiences have on those individuals. As with humans, dogs hold onto past traumas (more on this in Chapter 5), which can lead to behavioural problems, reminding us that welfare is not a momentary state but something an animal experiences throughout their lifetime.

There is also a need to consider the duration of positive and negative experiences and the intervals between them, as these can also have a lasting effect on welfare. This tool also assesses the animal's physical wellbeing, psychological wellbeing, environmental comfort, veterinary procedures, and management procedures. In practice, the AWAG is used to assess a dog's emotional state, behaviour and how it responds to various aspects of life.

Maslow's Hierarchy of Needs

A further framework for ensuring that welfare needs are met is Maslow's Hierarchy of Needs (Fig. 2.1). Though originally organized according to human needs, this analogy can be applied to all animals. Maslow theorized that physiological needs were the most fundamental, prioritizing basic survival (hence placed at the base of the pyramid). Once those needs were met, he suggested that humans would seek to meet needs related to safety, followed by love, self-esteem and eventually self-actualization.

Recent studies have sought to establish a hierarchy of needs applicable to dogs, providing a valuable framework for helping domestic dogs to thrive (Fig. 2.2). As with

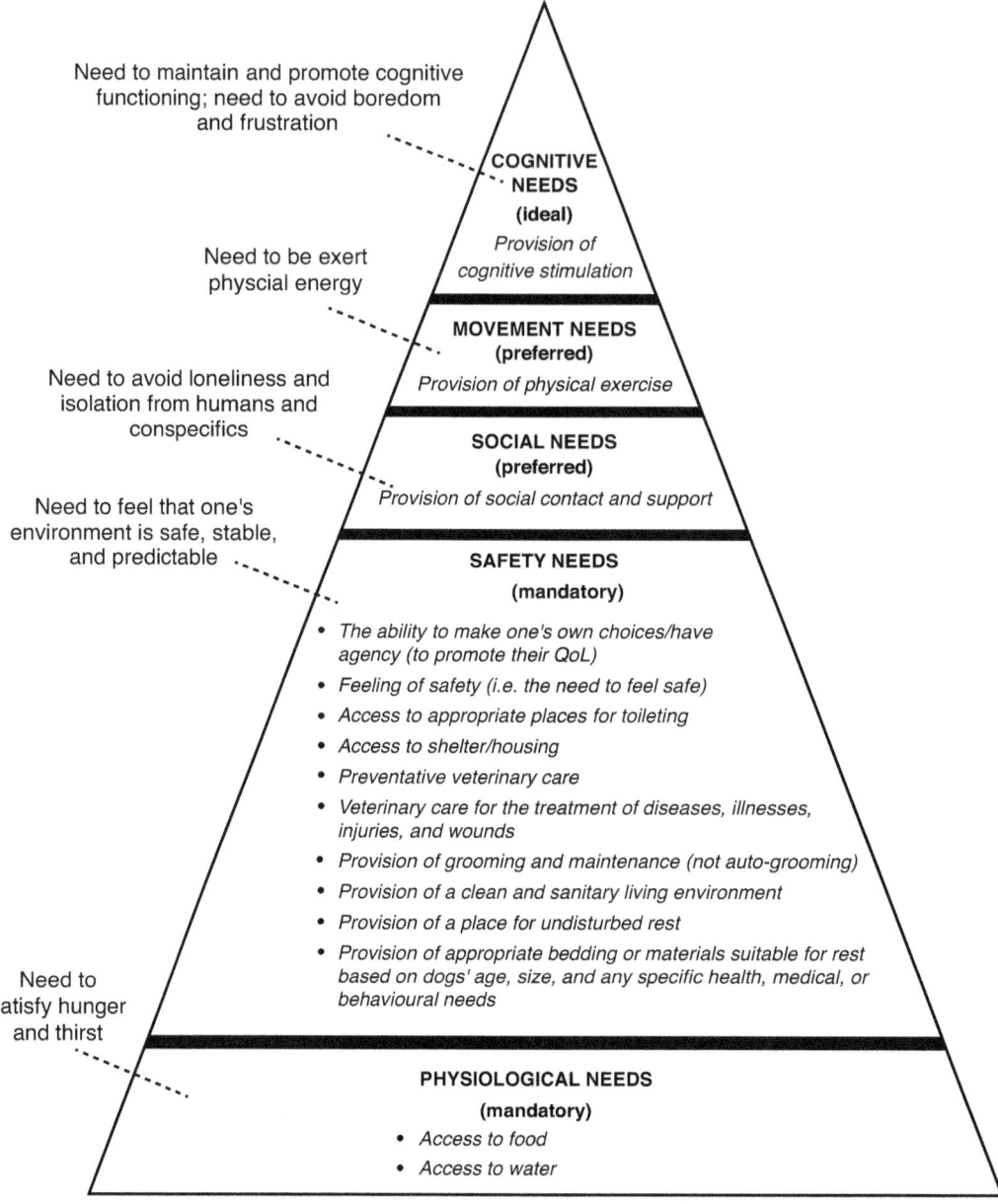

Fig. 2.2. Hierarchy of dog's needs (Griffin *et al.*, 2023). QoL, quality of life. Image is open access.

humans, dogs will prioritize physiological needs as well as safety. Given that they are a social species, a secure social group is also a basic requirement (though the demographic of such will likely differ from dog to dog). Finally, integrity and cognitive needs should be addressed as shown in the graphic (Griffin *et al.*, 2023), aligning with human need for esteem and self-actualization.

Essentially, if a dog's physiological needs are not met they cannot feel safe and so on. Maslow's Hierarchy provides caregivers and canine professionals with a basic framework to follow, and all of us should seek to be completing the pyramid.

Animal Ethics

As with ethical questions concerning human society and behaviour, the topic of animal ethics is hugely varied and divisive. Prioritizing the emotional wellbeing of a dog falls mostly in line with an Animal Rights ethical framework. An overview of Animal Rights theory is as follows:

> Agents [humans] and patients [dogs] are conscious, possess a complex awareness, and have a psychophysical identity over time. Agents and patients may be harmed or benefited and have a welfare in that their experiential life fares well or ill for them, independently of utility that they have for others or the interest that others have in them. Inherent value theory holds that the individual has a distinct moral value that is separate from any intrinsic values and that the attribution of equal inherent value to both moral agents and relevantly similar moral patients is required because both agents and patients are subjects-of-a-life.
>
> (Francione, 1997)

In lay terms, humans and dogs are both aware, conscious beings who have thoughts, feelings, and a sense of themselves over time. They can be hurt or helped, and their lives can go well or badly from their own perspective, not just based on how useful they are to others or how much others care about them. Each individual has worth just because they are a living, feeling being, not because of what they can do for others.

The ideas outlined in this book are not designed to make life easier for humans at the expense of their dog's emotional wellbeing. For this reason, *behaviour modification techniques will involve*:

- Fear-free techniques.
- Embracing natural behaviours.
- Consent-based methods.
- Consideration for the impact of pain and discomfort on behaviour and wellbeing.
- Human compromise.

Arguably however, it is not possible to completely align with an Animal Rights perspective where management is required for safety. This is just one of many ethical questions raised with regards to the keeping of animals. For example, Mason and Burn (2011) discuss the fact that even having a lack of control over one's own life is likely to be 'inherently stressful' for the animal. With this in mind, humans must do everything they can to provide dogs with as much choice, control and agency as possible in order to enrich their lives and boost their emotional wellbeing.

Conclusion

'Welfare' and 'emotional wellbeing' are complex concepts that have been the subject of much debate. The examples discussed move past the notion that welfare is something humans give to animals and instead ask us to think about what we can do to allow organic emotional wellbeing to occur. While our dogs cannot converse with us, they show us in their own way if welfare is poor by providing us with behavioural and physiological indicators, and it is important to remember that the deciding factors influencing their outlook will likely vary between individuals. Caregivers and practitioners need to be tuned in and ready to listen, both for the sake of the animal and for the fulfilment of their mutual relationship.

References

Animal Welfare Act (2006). Available at: https://www.legislation.gov.uk/ukpga/2006/45/contents (accessed 10 July 2025).
AWAG (2025) About AWAG. Available at: https://awag.org.uk/about (accessed 7 June 2025).
Broom, D.M. (1998) The scientific assessment of animal welfare. *Applied Animal Behaviour Science* 20, 5–19.
Francione, G.L. (1997) Animal rights theory and utilitarianism: Relativenormative guidance. Available at: https://www.animallaw.info/article/animal-rights-theory-and-utilitarianism-relative-normative-guidance (accessed 7 June 2025).
Fraser, D. (2009) Assessing animal welfare: Different philosophies, different scientific approaches. *Zoo Biology* 28, 507–518.
Griffin, K.E., Arndt, S.S. and Vinke, C.M. (2023) The adaptation of Maslow's Hierarchy of needs to the Hierarchy of Dogs' needs using a consensus building approach. *Animals* 13(16), 2620. Available at: https://doi.org/10.3390/ani13162620
Mason, G.J. and Burn, C.C. (2011) Behavioural restriction. In: Appelby, M.C. (ed.) *Animal Welfare*, 2nd edn. CABI, Wallingford, UK, pp. 98–119.
Office for National Statistics (2019) Measuring national well-being in the UK: International comparisons, 2019. Available at: https://www.ons.gov.uk/peoplepopulationandcommunity/wellbeing/articles/measuringnationalwellbeing/internationalcomparisons2019 (accessed 10 July 2025).
Ohl, F. and Putman, R. (2018) *The Biology and Management of Animal Welfare*. Whittles Publishing, Dunbeath, UK.
PDSA (2024) *PDSA Animal Wellbeing (PAW) Report*. Available at: https://www.pdsa.org.uk/what-we-do/pdsa-animal-wellbeing-report/paw-report-2024 (accessed 10 July 2025).

3 The History of Dogs

SOPHIE EAST[1] AND JADE NICHOLAS[2]*

[1]Brighton, UK; [2]CAB, Winchester, UK

Abstract

This chapter explores the process of domestication and its influence on dogs' thoughts, feelings, and behaviour, highlighting the key differences between domestic dogs and their wild ancestors. It also examines the lives of modern-day free-ranging or 'semi-feral' dogs to help readers consider how pet dogs might behave in alternative environments. By understanding dogs' evolutionary history and the experiences of their free-ranging counterparts, readers can gain deeper insight into the emotional wellbeing of their own dogs today.

Introduction

Dogs have lived alongside humans for thousands of years, evolving slowly into the companions we currently share our homes with. Through domestication, it is thought that wild grey wolves (*Canis lupus*) have transformed into the loving and loyal animals we know today (*Canis familiaris*). Sadly, many people around the world still try to treat domestic dogs as they would their wild ancestors, often leading to problems. For example, the frequently misunderstood 'dominance theory' still lingers from the 1940s (Schenkel, 1947), written based on observations on captive wolves. While unskilled practitioners may still use this theory to make up for gaps in their knowledge, modern science demonstrates the importance of treating the domestic dog in the present, understanding that they are different in many ways to their ancestors. Some scientists believe that domestic dogs are now so far removed from their ancestors that they should be '[their] own species within the taxonomic family called Canidae and the genus *Canis*' (Coppinger and Coppinger, 2016).

When referring to domestication, we refer to the process of rearing an animal or a group of animals in captive artificial habitats, selectively breeding them for human purposes, and controlling their reproduction and food supply to create a species dependent on humans for survival. In short, it is 'the adaptation of animals for human consumption and utilization' (Adhikari, 2023).

The domestication of animals can be seen throughout history, with species such as pigs, cows, and sheep being domesticated around 9500 BCE (Before the Common Era). Some research dates the domestication of dogs to nearly 35,000 years ago, although there is speculation around the exact timeline of canine domestication (Adhikari, 2023).

Failing to recognize the rich ancestral history of domestic dogs can cause humans to forget their biological needs, leading in turn to poor emotional wellbeing. Conversely, having a clearer understanding of canine domestication provides the answers to many common questions and dispels the idea that domestic dogs are wild animals consistently seeking 'alpha' position in the social group.

*Corresponding author: aboutyourdog@outlook.com

The Domestication of Dogs

The domestic dog belongs to the Canidae family, which consists of 35 related species that separated within the last ten million years. Thanks to advances in DNA sequencing, scientists have been able to partially reconstruct the dog's family tree. These advancements revealed wolf-like canids or grey wolves as the closest living 'cousins' of domesticated dogs. In fact, dogs share around 98% of their mitochondrial DNA with wolves (Marshall-Pescini and Kaminski, 2014).

However, research is ongoing.

Despite the genetic closeness, the exact origins of the domestic dog remain unclear and are still being studied. In 2015, a 35,000-year-old wolf bone was found below a frozen cliff in Siberia. The genetic information provided by this bone suggests that canine domestication happened much earlier than initially thought. However, this is speculative evidence, and most scientists theorize that canine domestication occurred between 15,000 and 30,000 years ago (Dunham, 2015).

While we may not know the exact timeline, we do know that dogs were among the first domesticated species, originating from grey wolves. Due to the expansive history of canine domestication, there is no single theory regarding how or why dogs came to be. Some theories suggest that wolves began living near human settlements and used them as a new food source, scavenging for scraps and human-generated waste. Other theories suggest that humans started capturing and breeding wolves for hunting purposes, forming a mutualistic relationship between both species (Galibert *et al.*, 2011). However, these theories are not strictly examples of domestication (Coppinger and Coppinger, 2016), as the wolves in these scenarios are not entirely reliant on humans for survival. Instead, they represent species that have benefitted from living in proximity to one another.

The true domestication process probably began when humans started selectively breeding wolves for specific traits, forming a much closer relationship between the two. The early stages of domestication focused on reducing fear and anxiety, producing a species much better suited for living in human environments. As this process continued, however, the focus shifted to breeding dogs that were better at forming social bonds and cooperation (Tancredi and Cardinali, 2023).

This transformation from wolf to dog was not only physical but also emotional. Dogs evolved in ways that make them uniquely responsive to human emotions. Their ability to recognize our moods, seek comfort, and form deep social bonds directly results from the long and ongoing journey of domestication.

It is important to remember that these are examples of theories about domestication that aim to explain the beginning of the process rather than its entire history. As well as this, domestication is not simply a moment or achievement in time, but an ongoing process (Losey, 2022). While selection in breeding contributes to the domestication process, it does not sustain it over multiple generations. Factors including care, mutual dependence, material resources, and shared spaces play key roles in sustaining and developing domestic relationships (Losey, 2022).

Changes in Genetics

From the moment the first dogs were separated from their wild ancestors, selective pressures began to shape their genetic composition, resulting in changes that influence both physical and behavioural traits. Understanding these changes allows for a greater comprehension of how the emotional needs of dogs have evolved over time.

One of the most significant differences between wolves and dogs is the dog's unique ability to form social bonds with humans, influenced largely by hormones in the brain such as glucocorticoids and oxytocin (Tonoike *et al.*, 2022).

Glucocorticoids

Glucocorticoids (GCs) such as cortisol and corticosterone are released during stress and high arousal situations. They help regulate metabolism and mobilization of energy, but they also affect behaviour, particularly responses to stress and threat.

High GC levels are typically observed during social conflict, competition or perceived danger. Such situations include fights, threats from rival groups or social instability. GC levels also rise during events involving separation anxiety, loss or social disruption.

For example, GC hormones in wolves are known to spike after the death of a dominant pack member or during the separation or disruption of pack members, reflecting the strength and complexity of wolf social bonds. In contrast, domestic dogs do not consistently show the same GC elevation when separated from other dogs; however, GC elevation is observed when domestic dogs are separated from their human companions. This suggests that domestication may have shifted the focus of dogs' social attachment primarily toward humans (Wirobski *et al.*, 2023).

Oxytocin

Oxytocin (OT) is released during positive social interactions and promotes further contact with a social partner. In humans, OT is linked to group activities that involve coordination. Higher levels of OT are associated with helping members of one's own group, but not outsiders. This hormone plays a crucial role in promoting loyalty and group conformity.

In animals, OT is released during positive social interactions. For example, OT levels increase in primates during social grooming and nurturing. In dogs, OT increases in response to positive interactions with humans, including petting, talking and mutual gaze (Miklosi, 2015).

As this hormone promotes loyalty, social memory and cooperative behaviour, dogs who displayed these traits were likely to be selected for breeding to improve human–dog cooperation.

Interestingly, male dogs tend to exhibit higher baseline OT levels compared to females, although this can vary based on context, breed and individual temperament. Female dogs often show OT levels that are more similar to wolves (Wirobski *et al.*, 2023). This may suggest different evolutionary pressures relating to nurturing and caregiving roles.

How have these genetic changes affected domestication?

Both oxytocin and glucocorticoids seem to play important roles in the domestication process. For example, in classic experiments with silver foxes, scientists selectively bred foxes that were less fearful and more friendly towards humans. After just a few generations, these now tame foxes had lower levels of stress hormones because their stress system became less active (Wirobski *et al.*, 2023).

However, this pattern is not always the same across species. Dogs and wolves do not always show consistent differences in their stress hormone levels, perhaps suggesting the story is more complex. Similarly, the link between canine domestication and oxytocin is complex. It is heavily suggested that OT plays a role in domestication, particularly in dogs. However, there have not been many studies directly comparing how dogs and wolves respond to human contact in terms of OT. There is considerable research, however, on how pet dogs and their owners respond to friendly interactions (like petting and eye contact). Many of these studies noted increased OT levels in both the dog and the human. It is important to note that not all studies find this effect, which means that OT levels may increase only in certain situations or with specific types of interactions (Wirobski *et al.*, 2023).

Behavioural Changes Through Domestication

There is no doubt that domestication influences behaviour, but it does not fully account for the human-compatible behaviours observed in domestic dogs. Research has shown that wolves do maintain the necessary biological and cognitive criteria for responsiveness to human gestures (Udell *et al.*, 2009). However, it is the domestic dog's willingness to follow human action and accept them as social companions that sets these two animals apart.

By accepting humans as social partners, domestic dogs have developed a sensitivity to human social cues. This sensitivity has allowed them to succeed at tasks requiring them to respond to human actions or instructions. In this section, the term 'sensitivity' is used to describe how dogs react when certain stimuli or signals are present. It is important to remember that dogs may notice signals without showing an obvious reaction (Udell *et al.*, 2009).

With this in mind, a dog is considered sensitive to human signals if it alters its behaviour in a way that helps it obtain reinforcement or reward dependent on the instruction of its human counterpart. This behaviour response likely contributed to the success of domesticated dogs in human environments (Udell *et al.*, 2009).

Udell *et al.* (2009) go on to provide some examples:

- Noticing if a person is paying attention.
- Learning the meaning of words.
- Copying what people do (social learning and imitation).
- Following where a person points.

Another factor contributing to dogs' willingness to accept humans as social partners is the timing of socialization. Dogs are often raised with humans from a young age and are therefore more likely to seek out human attention. Unless intensely socialized very early, wolves are less interested in human interaction. Dogs also possess a much larger window for when they can form social bonds, meaning that if not initially introduced to humans from birth, the likelihood of creating a social bond is still high. For wolves, this window is shorter and starts earlier in life. Studies suggest that this window of socialization can be as short as 3 weeks from birth. This longer 'sensitive period' in dogs suggests they are more likely to form strong bonds with people than wolves (Udell *et al.*, 2009).

So far, this chapter has explored the emotional aspects of domestication. However, physical changes can also influence behaviour. For instance, dogs have evolved into different breeds, exhibiting variations in coat colour, ear shape, fur type and reproductive differences. These traits can affect how dogs behave and interact with the world as they grow.

One common change in dogs is paedomorphosis, the retention of juvenile (puppy-like) traits into adulthood. This process can affect behaviour and communication and is often most noticeable in breeds that look very different to wolves. For example, breeds like Bulldogs or Cavaliers tend to display more puppy-like behaviour even through to adulthood, while breeds that closely resemble wolves, like huskies, are observed to behave more like adult wolves (Udell *et al.*, 2009).

The process of selective breeding has also affected how different breeds communicate and express themselves. High arousal, fear or aggression is typically communicated through the ears and fur. Some short-haired breeds raise their hackles during high-arousal situations, and some dogs demonstrate changes in ear positioning, for example ears flattened back often indicate fear. In contrast, breeds with floppy ears or a longer or curly coat probably exhibit these same behaviours, but their physical differences make it harder to notice signs of aggression or fear. This can limit the ways in which dogs send social signals (Udell *et al.*, 2009).

Modern Implications of Domestication

The domestic dog, along with the many breeds that form this diverse subspecies, is a growing area of interest in behavioural research. In the scientific community, there is an ongoing debate on how to define 'breed'. There are two main ways dog breeds are defined or categorized; each category reflects a different view of what is meant by 'breed'. (Mehrkam and Wynne, 2014).

A newer scientific approach groups breeds based on genetic similarities. For example, in 2005, researchers Parker and Ostrander studied the DNA of 85 dog breeds and grouped them into four genetic categories:

1. Asian and African breeds.
2. Mastiff-type dogs.
3. Herding dogs and sighthounds.
4. Modern hunting dogs.

Some breeds may belong to more than one of these groups. Parker and Ostrander's (2005) classification of breeds suggests that many dog breeds were recently created through crossbreeding, resulting in genetic similarities.

The more traditional method of defining 'breed' is grouping dogs based on the 'jobs' they were originally bred to do. Kennel clubs, such as the American Kennel Club (AKC), use this method. The AKC recognizes 180 breeds divided into seven groups.

Despite the differences in how breeds are grouped, scientists generally agree that dog breeds show consistent and predictable behaviour differences. Each breed tends to have a unique way of behaving, which is often referred to as the behavioural breed standard (Mehrkam and Wynne, 2014). Some breeds were selected for physical traits (like size), which influence their behaviour, while others were bred specifically for behavioural traits, such as herding instincts or guarding skills (Mehrkam and Wynne, 2014). These traits then become central to that breed's identity.

As mentioned in this chapter, domestication is not a singular achievement or event in history but an ongoing process. Behavioural breed standards will have changed over time due to cultural changes and the modernization of society (Svartberg, 2006). The selection and use of dogs are vastly different today compared to how they were traditionally used. The practical functions of dogs have been gradually diminishing, with the dog's role morphing into a companion and an object of affection.

Behavioural differences in modern domestic breeds can be displayed through variations in playfulness, curiosity, fearlessness, sociability and aggression. While behaviour variations are more common across breeds, dogs can also show individual variations within their breed. Interestingly, some patterns in behaviour variations are not linked to a breed's historical purpose but to how the breed is currently being used.

Free-Ranging Dogs

As well as the fascinating evolution and domestication of the grey wolf into our domestic dogs, around the world a population of semi-feral, free-ranging or 'street dogs' have stood the test of time. While the UK does not have a population of free-ranging dogs, it is common for countries in Europe, Africa, Asia and South America to have diverse populations in great numbers, overall accounting for an estimated 80% of the world dog population (Bhattacharjee et al., 2018). This book is by no means an in-depth look at these populations of dogs, but failing to provide a short overview of free-ranging dog behaviour potentially misses a core piece of the emotional wellbeing puzzle.

Free-ranging dogs are thought to have undergone a secondary process, their ancestors having started out as pet dogs (Miklosi, 2015). Now, their relationship with humans in these settings is complex, as they have evolved to understand that humans can both help and hurt them (Bhattacharjee et al., 2018). These dogs provide a unique look at what our domestic dogs might choose to do if they had total freedom and autonomy, further still illuminating where many modern-day behaviour problems might be coming from.

Dogs are opportunistic scavengers. The group of dogs that populate the Mexico City dump forage through the discarded rubbish in order to feed themselves (Coppinger and Coppinger, 2016), and similar traits exist in our pet dogs. Many pet dogs will scavenge (and are frequently chastised for doing so), and many dogs enjoy puzzles that require some element of problem solving to obtain reward. Though not necessarily preferable, research has found that pet dogs are happy to perform tasks in order to obtain food (Rothkoff et al., 2024) and anecdotally, it is not uncommon for them to show a preference for doing so. For this reason, it is particularly frustrating seeing dogs punished for counter-surfing or bin-raiding when they are biologically motivated to perform such behaviours.

Outside of the pet home, dogs often have complete freedom of movement, as well as the freedom to choose their social groups. They also have the freedom to choose how to spend their time, with studies revealing that the majority of their time is spent resting and sleeping (Jones, 2021). This observation raises concerns about the dogs in family homes that do not get opportunity to switch off or are regularly over-exercised. Later in the book a discussion is raised regarding dog day-care facilities, a primary concern being the 8 hours of hyper-arousal and activity that occurs when a dog attends for the day.

Chapter 5 discusses trauma in dogs imported from overseas (amongst other groups), and one of the suggested reasons for heightened stress levels in these dogs is due to lack of control over their situation. If free-ranging dogs are in the vicinity of a dog outside of their social group, they have the agency to be able to move away to somewhere safer or more comfortable. When kept on a lead, or even sharing a house with another dog, that agency is completely taken away, a highly stressful experience.

Overall, free-ranging dogs are frequently observed to be relaxed and passive individuals (Jones, 2021), and indeed, many behaviour professionals have found them to be happy and well-regulated. Of course, they face challenges that domestic dogs do not face, but anyone prioritising the emotional wellbeing of a domestic dog can learn a lot from the lifestyles chosen by free-ranging dogs.

Conclusion

Humans and domestic dogs have a long shared and complex history. The history of our relationship with dogs remains a slight mystery, but we do know it started with a mutualistic relationship between humans and wolves. This relationship then led to selective breeding for desired physical and behavioural traits. The result of this long and ongoing process is the domestic dog, an animal capable of forming deep social bonds and learning the nuances of a complex human world.

However, this process does not come without its setbacks. The development of breeds through selective breeding raises numerous physiological and behavioural concerns. Breed-specific behaviour may become misunderstood or labelled as problematic owing to misconceptions or misunderstandings about the dog's function. Furthermore, this move away from function and growing focus on aesthetics may worsen the physical and mental health and wellbeing of modern domestic dogs.

Due to the extensive history of dogs and their ongoing development, it is important for both caregivers and practitioners to familiarize themselves with the ancestral past of the domestic dog, understanding of course that they have changed significantly since humans first domesticated wolves. By comprehending their original function and biological history, as well as the behaviour of their free-ranging counterparts, humans may be better equipped to care for them and maintain high welfare standards.

References

Adhikari, R.C. (2023) Domestication of animals. *Medha, A Multidisciplinary Peer Reviewed Research Journal* 6(1), 55–61.

Bhattacharjee, D., Sau, S. and Bhadra, A. (2018) Free-ranging dogs understand human intentions and adjust their behavioral responses accordingly. *Frontiers in Ecology and Evolution* 6. DOI: 10.3389/fevo.2018.00232.

Coppinger, R. and Coppinger, L. (2016) *What is a Dog?* University of Chicago Press, Chicago, Illinois.

Dunham, W. (2015) Dog domestication much older than previously known. *Scientific American*. Available at: https://www.scientificamerican.com/article/dog-domestication-much-older-than-previously-known/#:~:text=Previous%20research%20based%20on%20genetic,mutations%20occurred%20across%20the%20genome (accessed 10 July 2025).

Galibert, F., Quignon, P., Hitte, C. and André, C. (2011) Toward understanding dog evolutionary and domestication history. *Comptes Rendus - Biologies* 334, 190–196. DOI: 10.1016/j.crvi.2010.12.011.

Jones, E. (2021) What can "streeties" teach us about companion dogs? *The IAABC Foundation Journal* 19. DOI: 10.55736/iaabcfj19.5.

Losey, R.J. (2022) Domestication is not an ancient moment of selection for prosociality: Insights from dogs and modern humans. *Journal of Social Archaeology* 22(2), 131–148. DOI: 10.1177/14696053211055475.

Marshall-Pescini, S. and Kaminski, J. (2014) *The Social Dog*. Academic Press, San Diego, California.

Mehrkam, L.R. and Wynne, C.D.L. (2014) Behavioral differences among breeds of domestic dogs (*Canis lupus familiaris*): Current status of the science. *Applied Animal Behaviour Science* 155, 12–27. DOI: 10.1016/j.applanim.2014.03.005.

Miklosi, A. (2015) *Dog Behaviour, Evolution and Cognition*, 2nd edn. Oxford University Press, Oxford, UK.

Parker, H.G. and Ostrander, E.A. (2005) Canine genomics and genetics: Running with the pack. *PLoS Genetics* 1(5), 507–513. DOI: 10.1371/journal.pgen.0010058.

Rothkoff, L., Feng, L. and Byosiere, S.E. (2024) Domestic pet dogs (*Canis lupus* familiaris) do not show a preference to contrafreeload, but are willing. *Scientific Reports* 14(1), 1314. DOI: 10.1038/s41598-024-51663-x.

Schenkel, R. (1947) *Expression Studies on Wolves*. Zoological Institute of the University of Basle, Switzerland.

Svartberg, K. (2006) Breed-typical behaviour in dogs - Historical remnants or recent constructs? *Applied Animal Behaviour Science* 96(3–4), 293–313. DOI: 10.1016/j.applanim.2005.06.014.

Tancredi, D. and Cardinali, I. (2023) Being a dog: A review of the domestication process. *Genes* 14, 992. DOI: 10.3390/genes14050992.

Tonoike, A., Otaki, K., Terauchi, G., Ogawa, M., Katayama, M. *et al.* (2022) Identification of genes associated with human-canine communication in canine evolution. *Scientific Reports* 12(1), 6950. DOI: 10.1038/s41598-022-11130-x.

Udell, M.A.R., Dorey, N.R. and Wynne, C.D.L. (2010) What did domestication do to dogs? A new account of dogs' sensitivity to human actions. *Biological Reviews* 85(2), 327–345.

Wirobski, G., Range, F., Graat, E.A.M., Palme, R., Deschner, T. *et al.* (2023) Similar behavioral but different endocrine responses to conspecific interactions in hand-raised wolves and dogs. *IScience* 26(2), 105978. DOI: 10.1016/j.isci.2023.105978.

4 Emotional Wellbeing Starts with the Right Dog

JADE NICHOLAS*

CAB, Winchester, UK

Abstract

Of course, emotional wellbeing is largely about what choices we make for our dogs on a regular basis. However, many problems routed in poor wellbeing can be prevented at the start of our journey by making informed decisions and choosing an emotionally resilient dog. This chapter explores ways in which dogs can be selected for their robust and capable personalities in order to avoid problems caused by poor genetics, early trauma and irresponsible breeding.

Introduction

It is commonly said that there are 'no bad dogs, only bad owners'. While of course the behaviour of a caregiver is extremely pertinent in determining the behavioural outcome for their dog, this popular saying overlooks factors that were determined way before the dog belonged to them. For example, genetics, breed characteristics and learned traits all contribute to the temperament and behavioural outcome of the individual. It is therefore the responsibility of all humans involved (breeders, caregivers and consulting professionals) to set that dog up to have fantastic emotional wellbeing from conception through to end-of-life care. This chapter is about making informed choices and choosing an emotionally robust dog. When humans take the time to select for a healthy dog (both mentally and physically), the journey to emotional wellbeing is much easier, as the caregiver can enjoy a rich and varied existence with their canine companion. When dogs are purchased on impulse or the necessary research is not completed, however, caregivers will probably find themselves swimming against the tide.

Though not all caregivers will choose to obtain their dog from a breeder, this still remains the most popular choice (when considering recent research). It also remains a massively unregulated pathway compared to the rigorous rehoming systems run by charities like Dogs Trust and the RSPCA. Given the popularity of this option, a huge number of caregivers can make decisions benefiting both themselves and their dogs before the puppies even leave their mothers.

Where animals in rescue are concerned, there are often many unanswered questions about the history of that individual. Therefore, the emotional resilience of the dog is largely out of control of the new caregiver. However, reputable rescues will often help caregivers to choose breeds (or crossbreeds) based on their lifestyle and preferences, which is still a particularly important consideration for longevity in the family.

---

*Corresponding author: aboutyourdog@outlook.com

Choosing a Dog Breeder

In a survey of 2371 dog owners in the UK (PDSA, 2024) 33% of respondents shared that they had obtained their dog from a breeder, meaning that private breeding still has the potential to bring enormous welfare implications upon the UK dog demographic. The Dog Breeding Licence (England) requires breeders producing a certain number of puppies to obtain a licence. However, this leaves much to be desired in terms of 'hobby' breeders that breed below the required amount. For these individuals, a licence is not required, and so multiple welfare implications are called into question. As discussed in Chapter 5, puppy farming continues to generate dogs for the UK market (as well as in many other places) and so prospective caregivers are often at the mercy of the dog breeder in question.

In a 2024 survey by Brand, *et al.*, a number of dogs were reported to be relinquished before 21 months of age (n=4369). When asked, respondents commented that more pre-purchase support could have prevented the relinquishment of their dog, demonstrating how important an ethical and honest breeder can be in setting a dog up for success.

Scarily, the 2023 PDSA Paw Report found that only 43% of caregivers knew that they should view their puppy with its mum before purchase (PDSA, 2023). The Puppy Contract, a vital tool for protecting caregivers from unethical practice, was familiar to only 15% of caregivers, though this did increase in the 2024 report (23%). These two basic practices are certainly not 'fool proof', but they do go some way to ensuring that puppies are obtained from a reputable source with the correct protections in place. Otherwise, it is easy to sell puppies online and scam owners into the illegal puppy trade (Menor-Campos, 2024).

As well as inherited traits from the breeding pair, the breeding environment is known to have an impact on the litter. In particular, pregnant animals exposed to stressors can bear puppies with increased sensitivity to stress, impaired learning abilities and 'atypical social behaviours' (Menor-Campos, 2024). Social isolation can also lead to these outcomes.

An alarming trend in breeding is the increase in 'canine fertility clinics' (Loeb and Evans, 2020), making it easier for breeders to have litters using procedures such as artificial insemination and caesareans. Likely related to this new service, social media shows a lucrative new market for exotic dog breeds, colour-ways and coat types. For example, the popular French Bulldog is traditionally only brindle, fawn and pied in coat colour (Kennel Club, 2025). As a result of mass experimental breeding these dogs can be found in colours such as lilac, blue, chocolate and others. Social media breeders will boast about the exciting new coat colour they have found in their dog, and sell dogs carrying such genes at extortionate fees to other prospective hobbyists. At the time of writing this (2025), French Bulldogs are now produced both hairless and fluffy as well as their traditional smooth shorthaired coat. Not only this, but French and English bulldogs are available for stud at a huge proportion of these clinics, despite being breeds that have an 80% caesarean rate, unable to give birth naturally (Loeb and Evans, 2020).

The following should be 'red flags' when choosing a reputable breeder:

- Breeding for extreme conformation, particularly in breeds already compromised (a 'Pocket Bully' is a great example of this).
- Boasting new or innovative coat types and colours.
- Use of fertility clinics as routine, rather than in emergency.
- Advertising dogs as carrying certain genes (particularly coat colours and types).

The examples given above (along with many others), demonstrate a breeder whose only intention is to generate income and produce a portfolio of aesthetics worthy of social media. When questioned, a breeder should be prioritizing the physical health of the breeding pair as well as the temperament. The aesthetic of the dog should be a tertiary consideration to both the breeder and potential caregiver.

Rescue and Rehoming

Choosing a dog from a rescue organization or private rehome is likely to limit the information available to the caregiver. For example, it is not uncommon for dogs to have been passed around multiple times, meaning that adopters do not have access to information about their lineage or early experiences. Despite this, some important information will still be available and will ultimately help to make an informed decision.

As discussed in further detail below, the breed (or mix of breeds) might still be known and will still provide valuable information about potential health issues and typical behaviour characteristics. If there is some information about previous life experience along with temperament testing on site this should all be taken into account.

Animals imported from overseas, rescued from puppy mills and seized from criminal activity are all discussed in Chapter 5. Traumatized dogs require an accelerated level of care and commitment, but sadly they are frequently placed into the care of people who mean well but do not have the necessary knowledge or skillset. Inadequate 'pop up rescues' unable to provide suitable home checks, behavioural assessment or rescue back up for these dogs should be avoided at all costs. At best, these organizations are underprepared for dealing with the complex web of rehoming animals; at worst, they are fronts for animal cruelty that defies comprehension. Recently in the UK, an individual claiming to run a rescue and rehoming service was charged with animal cruelty and fraud offences after 37 missing dogs were found dead on his property (ITV News, 2025).

Potential caregivers can find a reputable rehoming organization via the Association of Cats and Dogs Home (ACDH).

The 'Right' Breed

As discussed throughout this book, physical and emotional wellbeing go hand in hand and cannot be passed over when it comes to enhancing the positive worldview of a dog. Therefore, a crucial decision can be made even before a caregiver meets a dog: the decision of which breed (or crossbreed) to invite into their home.

Physical considerations

It is well known that brachycephalic (flat faced) breeds are some of the most physically unhealthy dogs in the world (Morel *et al.*, 2024). However, French Bulldogs, Pugs and Shih Tzus remain some of the most popular breeds with humans, in particular contributing to a huge proportion of online advertisements. Dachshunds, often miniature, are pre-disposed to a lot of painful conditions, intervertebral disc disease (IVDD) to name but one (Packer *et al.*, 2016; Morel *et al.*, 2024), and yet ranked 4th in a survey of the UK dog-owner population (Dogs Trust, 2024).

Physical abnormalities are not limited though to small breeds, and many large breeds experience chronic joint problems (Bovenkerk and Nijland, 2017) amongst various other ailments. Border Collies, for one, ranked 3rd in the 2024 Dogs Trust survey are known to 'have at least 25 inherited disorders' (Menor-Campos, 2024). The trend in unhealthy dogs is expected to worsen if humans continue to prioritize aesthetics above all else.

It is sometimes said that choosing a crossbreed dog increases the likelihood of having a healthier pet. This, however, is not guaranteed. Many genetic changes in dogs and cats are linked to diseases, no matter what breed they are. Some of these disease-related genes have been around for a long time and can be found in both purebred and mixed-breed dogs; however, purebred dogs are more likely to inherit two copies of a harmful gene (one from each parent), making them more likely to get the disease. This happens because purebred dogs are often bred from a small group of ancestors, which reduces genetic variety. Mixed-breed dogs can also carry harmful genes, but it depends on which breeds are in their background, since some harmful genes appear in several breeds (Morel *et al.*, 2024).

Recommendations for potential caregivers (whether buying or adopting) should therefore consider the impact of physical health on emotional wellbeing. Breeders should be dissuaded from continuing lines of physically compromised dogs, since the popularity of certain breeds is still incredibly high. For example, a 2020 article shared that French Bulldogs have seen a 3000% increase in popularity between the years of 2009 and 2018 (Loeb and Evans, 2020). It is also important to consider whether a dog is healthy, or just healthy in terms of that breed. In a survey of 2186 brachycephalic dogs, 70.9% of caregivers reported their dog as healthy or 'in the best health possible' (Packer *et al.*, 2019), a frighteningly low bar for what is acceptable.

Behavioural implications

Though all dogs are individuals and will express themselves based on a combination of factors, there are some reliable predictors that can make a dog more likely to be emotionally robust from a young age.

One of these factors is understanding the characteristics associated with a certain breed and considering whether the potential caregiver is a suitable candidate to meet their needs.

'Working' breeds (those traditionally bred for work) are popular, as they have been shown to be marginally easier to train, more playful and less fearful (Asp *et al.*, 2015). However, with 19% of UK owners leaving their dog for 5 hours or more (PDSA, 2024), these alert and active breeds are frequently mismatched to their environment and behaviour complaints are the result.

> The well-being of an animal inherently unable to adapt to the environment in which it is placed may be compromised by both their own stress and that of their owner. Anecdotally there is a fashion amongst pet owners of some breeds to seek puppies from working lines, without considering the potential for extremes of working-related behavior they would find difficult to manage.
>
> (Morel *et al.*, 2024)

Border collies, for example, are often bred from working stock on farms. These highly intelligent dogs not only display natural herding instincts, which are crucial for farm life, but they can also learn and associate words as 'referential nouns' – words that represent specific objects. In one study, a border collie was able to learn the names of 1022 different toys (Pilley and Reid, 2011). Collies from farms placed into urban families frequently encounter behavioural difficulties, possibly due to their high intelligence and immaculate attention to detail. It is not unusual to see collies referred to behaviourists for 'herding' traffic, family members or other pets.

Some studies present evidence that certain breeds are more likely to exhibit aggression than others. For example, Chihuahuas and Dachsunds both scored higher than average for aggressive behaviour towards both humans and dogs (Duffy *et al.*, 2008), and both breeds were recorded in the same study as being more likely to exhibit serious aggression including bites or bite attempts towards people. This appears to be a noticeable trend across many small breeds (Menor-Campos, 2024). Given both breeds mentioned above are very popular, as well as a variety of other small breeds, it is reasonable to assume that these data are not accessible enough to the public, or that breeders are not doing enough to set their dogs and owners up for success.

All of this said, evidence suggests that there is much more to a dog's behavioural presentation and temperament than just breed (Morrill *et al.*, 2022), and that an individual is a product of both nature *and* nurture. But choosing the right breed is a good place to start.

Natural behaviours

Many behaviour problems can be managed or even avoided by caregivers and professionals having the correct level of understanding about the dog in front of them and what they were bred to do. For example, Huskies in urban environments are often walked on tight restrictive equipment in an attempt to stop them from pulling forward, a behaviour they were genetically selected to perform. German Shepherds are targeted with painful prong and shock collars to stop them from barking at visitors, being punished for following their guardian instincts. Spaniels are denied off-lead time lest they chase a bird while out in the field, and then they come home and are chastised for parading tea-towels and socks in an effort to 'flush' as they were bred to do.

Whether purchasing from a breeder or adopting from a rehoming centre, caregivers must understand what their dog was designed for (Fig. 4.1). This fundamental but so often overlooked consideration will help individuals to set their household up for success. Each breed carries common characteristics, and these characteristics allow caregivers to select a companion that will fit into their lives (Morel *et al.*, 2024). For example, considerations might include the physical and mental requirements of that breed; can the caregiver provide the necessary levels of engagement and activity? Further still, is the caregiver choosing a dog with the correct personality in terms of social engagement? Or are they setting a solitary and aloof breed up to fail by bringing it into a busy household?

It is also the responsibility of any professionals consulting with the dog to understand the history of the breed (or combination of breeds) in front of them. A Bloodhound primarily motivated by scent will not respond the same way to loose-lead walking as a Labrador. While both breeds learn the same way, their motivation is completely different.

Fig. 4.1. Rufus swims for a hunting dummy. Image is author's own.

Conclusion

Though this chapter does not provide a formula for selecting the 'perfect puppy', or 'perfect' rescue dog, both in terms of breed choice and method of procurement, it should demonstrate the importance of research and preparation.

Caregivers might like to ask themselves questions pertaining to the following:
Breeder (if applicable):

- Do I have confidence in this breeder?
- Has the breeder provided a puppy contract?
- Have I met the mother of the litter?
- What do I know about the father of the litter?
- Has the breeder prioritized health and temperament over aesthetic?

Rescue centre (if applicable):

- Do they have a return-to-kennels policy if things do not work out?
- Have I seen where the dog lives?
- What do I know about the dog's life before me?
- What advice has the centre given? Is it ethical and up to date?

Size:

- Is my house big enough?
- Have I seen an example of this breed at their full size?
- Can I hold onto this dog on a lead?
- Can I afford to feed and medicate (if needed) this dog?

Activity level:

- How much mental and physical stimulation does this dog require?
- Do I have the time, energy and knowledge to meet these needs?

Health:

- Am I aware of the potential health problems for this dog?
- Am I in a position to provide veterinary treatment for those problems?
- Can I afford to insure this dog?

With these questions in mind, it should be easier for potential caregivers to set their dogs up for success and optimum emotional wellbeing, though of course there are never any guarantees.

References

Asp, H.E., Fikse, W.F., Nilsson, K. and Strandberg, E. (2015) Breed differences in everyday behaviour of dogs. *Applied Animal Behaviour Science* 169, 69–77.

Bovenkerk, B. and Nijland, H.J. (2017) The pedigree dog breeding debate in ethics and practice: Beyond welfare arguments. *Journal of Agricultural and Environmental Ethics* 30(3), 387–412. DOI: 10.1007/s10806-017-9673-8.

Brand, C.L., O'Neill, D.G., Belshaw, Z., Pegram, C.L., Stevens, K.B. et al. (2022) Pandemic puppies: Demographic characteristics, health and early life experiences of puppies acquired during the 2020 phase of the COVID-19 pandemic in the UK. *Animals* 12(5), 629. DOI: 10.3390/ani12050629.

Dogs Trust (2024) Welcome to the results of the National Dog Survey 2024. Available at: https://www.dogstrust.org.uk/downloads/Dogs_Trust_NDS_Report_2024__.pdf (accessed 10 July 2025).

Duffy, D.L., Hsu, Y. and Serpell, J.A. (2008) Breed differences in canine aggression. *Applied Animal Behaviour Science* 114, 441–460.

Loeb, J. and Evans, E. (2020) Puppy power: Fertility clinics on the rise. *The Veterinary Record* 186, 140.

Menor-Campos, D.J. (2024) Ethical concerns about fashionable dog breeding. *Animals* 14(5), 756. DOI: 10.3390/ani14050756.

Morel, E., Malineau, L., Venet, C., Gaillard, V. and Péron, F. (2024) Prioritization of appearance over health and temperament is detrimental to the welfare of purebred dogs and cats. *Animals* 14(7), 1003. DOI: 10.3390/ani14071003.

Morrill, K., Hekman, J., Li, X., McClure, J., Logan, B. et al. (2022) Ancestry-inclusive dog genomics challenges popular breed stereotypes. *Science* 376(6592). https://doi.org/10.1126/science.abk063

News, I.T.V. (2025) Billericay man charged after remains of 37 dogs found during Essex Police raid. Available at: https://www.itv.com/news/anglia/2025-05-15/man-charged-after-remains-of-37-dogs-found-during-police-raid (accessed 11 July 2025).

Packer, R.M.A., Seath, I.J., O'Neill, D.G., de Decker, S. and Volk, H.A. (2016) DachsLife 2015: An investigation of lifestyle associations with the risk of intervertebral disc disease in Dachshunds. *Canine Genetics and Epidemiology* 3(1), 8. DOI: 10.1186/s40575-016-0039-8.

Packer, R.M.A., O'Neill, D.G., Fletcher, F. and Farnworth, M.J. (2019) Great expectations, inconvenient truths, and the paradoxes of the dog-owner relationship for owners of brachycephalic dogs. *PLOS One* 14(7), e0219918.

PDSA (2023) *PDSA Animal Wellbeing (PAW) Report*. Available at: https://www.pdsa.org.uk/what-we-do/pdsa-animal-wellbeing-report/paw-report-2023 (accessed 10 July 2025).

PDSA (2024) *PDSA Animal Wellbeing (PAW) Report*. Available at: https://www.pdsa.org.uk/what-we-do/pdsa-animal-wellbeing-report/paw-report-2024 (accessed 10 July 2025).

Pilley, J.W. and Reid, A.K. (2011) Border collie comprehends object names as verbal referents. *Behavioural Processes* 86(2), 184–195.

The Kennel Club (2025) French Bulldog. Available at: https://www.thekennelclub.org.uk/breed-standards/utility/french-bulldog/ (accessed 20 March 2020).

5 Traumatized Dogs

JADE NICHOLAS*

CAB, Winchester, UK

Abstract

It is not unusual for previously traumatized dogs to be rehomed to family homes. This chapter explains the impact that trauma has on a dog's worldview and what this means when they are trying to adjust to a new life in a new environment. It has previously been suggested that traumatized dogs might benefit from trauma-informed care (TIC), normally used when working with children. TIC covers three pillars: safety, connections and managing arousal and readers will learn techniques to meet these needs. These techniques are particularly appropriate for overseas rescues, puppy mill survivors and dogs exposed to criminal activity to name a few groups.

Trauma in Dogs

> The most pervasive and far reaching impact of complex trauma is the dysregulation of emotions and impulses.
>
> (Bath, 2008)

Many dogs come from a situation or set of circumstances that we would describe as 'traumatic'. They may have experienced trauma in a single event such as a car accident (acute trauma), or others may have had experienced continued exposure to trauma (complex trauma). Common examples of the latter include dogs used in puppy farming or those who have been transported from overseas. Some charities rehome dogs from slaughterhouses and meat markets, perhaps the worst trauma imaginable.

Behaviourists and veterinary professionals should be proficient in working with dogs who have experienced trauma. Similarly, caregivers should be fully informed about the responsibility they are taking on when they choose to share their life with a traumatized animal.

Recent studies have compared trauma experience and recovery in humans and dogs (Corridan *et al.*, 2024). Research into the experience of children shows that trauma (especially in very early life) causes a permanent change in the brain, prompting the seeking of safety above all else rather than exploration and activity (Bath, 2008). This demonstrates how important it is to treat the patient in front of you as an individual, particularly if they have a history consistent with exposure to trauma. Failure to recognize their altered perception of the world around them could easily lead to inefficient treatment.

Not only is untreated trauma problematic from an emotional wellbeing perspective, the behaviour of traumatized animals can massively compromise their future prospects. A common behavioural consequence of exposure to trauma is extreme fearfulness, which can present in a multitude of ways. Ultimately, these dogs are said to have a reduced

---

*Corresponding author: aboutyourdog@outlook.com

quality of life and if they are in a rescue facility awaiting adoption they are less likely to be successful (Collins *et al.*, 2022).

Examples of behaviours or characteristics shown by traumatized dogs might include:

- Aggression to people or animals.
- Excessive vocalization.
- Separation related problems and insecure attachments to people or other animals.
- Fear of novelty and generally low confidence.
- Fear or phobia of car travel.
- Self-soothing behaviours such as excessive licking or chewing.
- Other abnormal repetitive behaviours such as light chasing or spinning.

It is suspected that there are many similarities regarding how people (particularly children) and dogs behave after trauma (Corridan *et al.*, 2024). Given the various similarities, researchers have suggested adopting a 'trauma-informed care' (TIC) approach when treating these individuals, based on principals that have been utilized for many years in human psychology. It's incredibly important to treat these cases with the necessary gravity in order to prioritize the emotional wellbeing of the animal.

> Trauma produces actual physiological changes, including a recalibration of the brain's alarm system, an increase in stress hormone activity, and alterations in the system that filters relevant information from irrelevant. We now know that trauma compromises the brain area that communicates the physical, embodied feeling of being alive. These changes explain why traumatized individuals become hypervigilant to threat at the expense of spontaneously engaging in their day-to-day lives. They also help us to understand why traumatized people so often keep repeating the same problems and have such trouble learning from experience. We now know that their behaviours are not the result of moral failings or signs of lack of willpower or bad character - they are caused by actual changes in the brain.
>
> (van der Kolk, 2014)

General Trauma-Informed Care Principles

The 'Three Pillars of Trauma-Informed Care' (Bath, 2008) could be a useful framework for treating trauma in animals (Fig. 5.1).

Based on extensive research in traumatized children, the below pillars are said to be 'fundamental and universal':

1. Safety.
2. Connections.
3. Managing emotions.

Considerations based on these three pillars in TIC should be applied to all canine trauma cases. Proposed measures to be implemented by caregivers (and other custodians of the dog) might include the following.

Safety

- Ensuring dogs have access to safe space (this can also be a safe person or persons when away from the home environment).

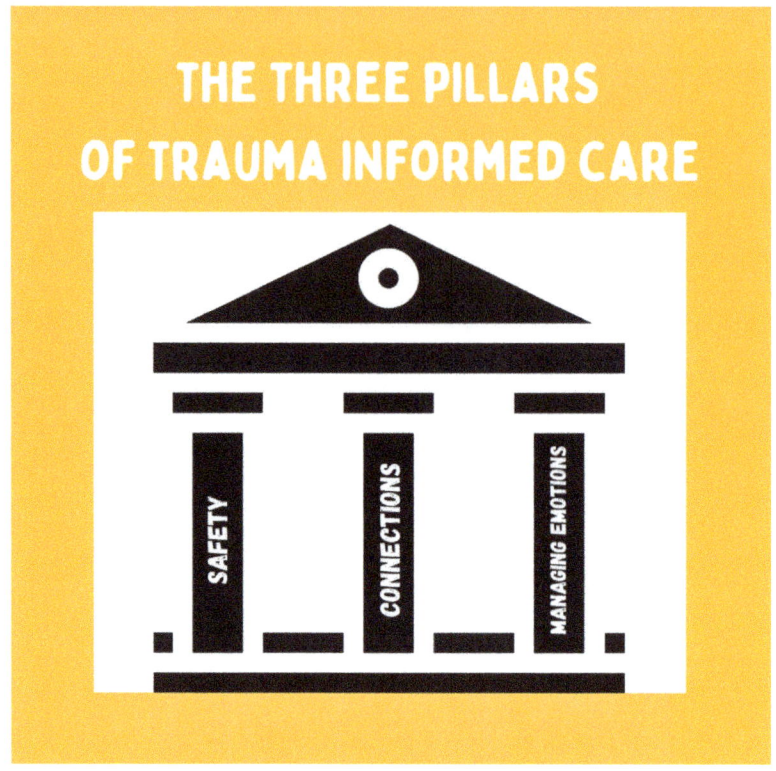

Fig. 5.1. The 'Three Pillars of Trauma-Informed Care', as written in Bath's 2008 paper. Image is author's own, made using Bath's original ideas.

- Offering predictability and continuity in routine and everyday occurrences.
- Involving dogs in decision making.
- Adopting consent-based protocols.

Connections

- Using only ethical training and behaviour modification practices, avoiding punishment or any behaviours likely to create fear.
- Ensuring that relationships in the home are healthy and positive, including those with other animals.

Managing emotions

- Teaching impulse control exercises.
- Teaching calm desirable behaviours as alternatives to dysfunctional behaviours.
- Providing suitable opportunities for decompression and the regulation of arousal.
- investigating the underlying problem leading to a behaviour problem (akin to listening to children when they are worried about something).

Guidance on how to apply such measures will follow later in the chapter.

As previously mentioned, there are many reasons a dog might be deemed to have experienced trauma. Acute trauma with lasting effects can happen to healthy and happy dogs in pet homes without warning. However, certain groups are more likely to have been exposed to trauma in large numbers. For the purposes of this chapter, the examples given will be overseas rescues (otherwise known as free-ranging or 'street' dogs), puppy mill survivors and those impacted by breed specific legislation and/or criminal activity. While these groups are the primary focus, the research and subsequent guidance given can be applied to any traumatized dog.

Overseas Rescue Dogs

The 2024 PDSA Paw Report found that 5% of all dogs in the UK (640,000 dogs) had come from overseas, an increase from 3% in 2020 (PDSA, 2024). Most commonly, dogs are imported from Romania as well as Spain, Cyprus and Ireland, though many other countries import dogs to the UK. Fig. 5.2 from the same survey details why dog owners chose to go down this route to acquire their dogs.

Interestingly, more than 10% of caregivers have chosen to rescue from overseas after being turned down by rescue centres at home. From a welfare perspective, this raises concerns as such owners might ask themselves why they were rejected in the first place. Perhaps they are unable to meet welfare standards in the UK but overseas organizations are more relaxed.

Anecdotally, as a behaviourist working in this field, it is common to see dogs from overseas rehomed to completely unsuitable environments. For example, it is not unusual for these dogs to be placed with first-time dog owners, in small flats in busy cities, or with children who are too young to safely be around a dog whose history is unknown. These observations raise enormous concerns about human and animal wellbeing as well as safety.

For those caregivers who do research their new companion and want to work to integrate them into their home, it is important to provide as much scientific evidence as possible to support them. Given the large numbers of dogs joining 'pet dog' homes it would be negligent not to explore the emotional wellbeing of these individuals.

One thing caregivers and practitioners can do to understand how to work with dogs from overseas is to understand the behaviour of free-ranging dogs, as discussed in Chapter 3.

Behaviour and Welfare Concerns in Dogs Imported From Overseas

Limited data have been collected on the behaviour of imported dogs in pet homes up to this point, but there are some studies of significance. A 2020 study (Norman *et al.*, 2020) collected survey responses from 3080 owners of dogs from overseas. One of the main aims of this study was to gather data on the prevalence of behaviour problems in these dogs, the results of which can be seen in Fig. 5.3.

Data collected in the 2020 survey correlate with other similar studies. For example, Munkeboe *et al.* (2021) compared the prevalence of behaviour problems in imported street dogs with those in puppies reared in Denmark. Both of the studies cited found fear responses and aggressive behaviour to be notable, with the Danish study reporting that imported dogs were significantly more likely to be scared of people than their pet counterparts (Fig. 5.4). Other sources note that imported dogs are more likely to be described as

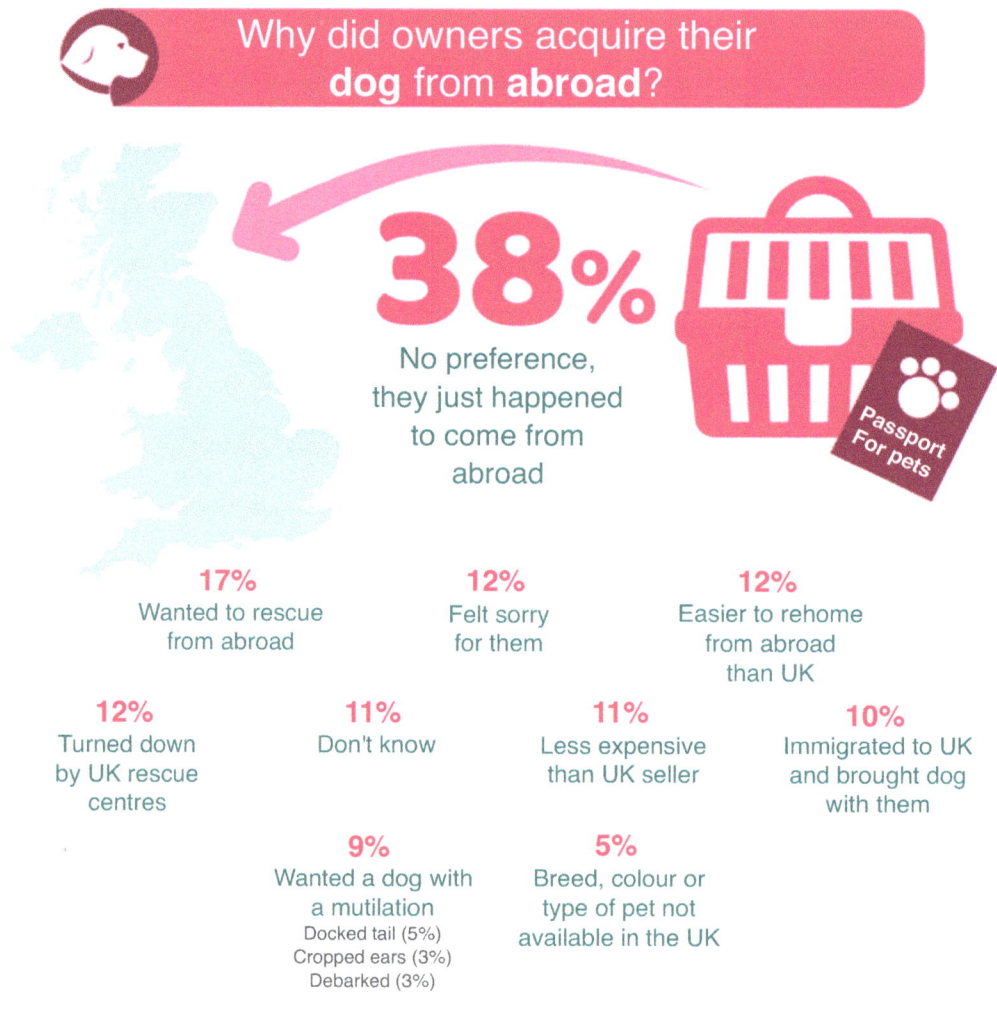

Fig. 5.2. Why did owners acquire their dog from abroad (PDSA, 2024)? Used with permission.

'stressed' and that they are less likely to be training or goal oriented (Zeiman, 2015). This finding is particularly interesting when comparing these dogs to the lack of task-driven motivation in traumatized children described earlier in the chapter.

74.7% of respondents ($n=75$) in a 2014 study reported that their dog was timid or fearful when *first introduced* to their new home (Salgirli Demirbas *et al.*, 2014). However, the same respondents reported that the majority of free-ranging dogs rehomed to pet homes had adapted positively *over time* (typically 6 months), demonstrating that many dogs are resilient enough to settle eventually with the right support structures in place.

Considering how common a fear of people and of novelty has proven to be in this small summary of research it is hardly surprising that many dogs are traumatized and struggling to cope once rehomed to a suburban life with humans. Understandably, some individuals and organizations are strongly against the relocation of street dogs from their homes.

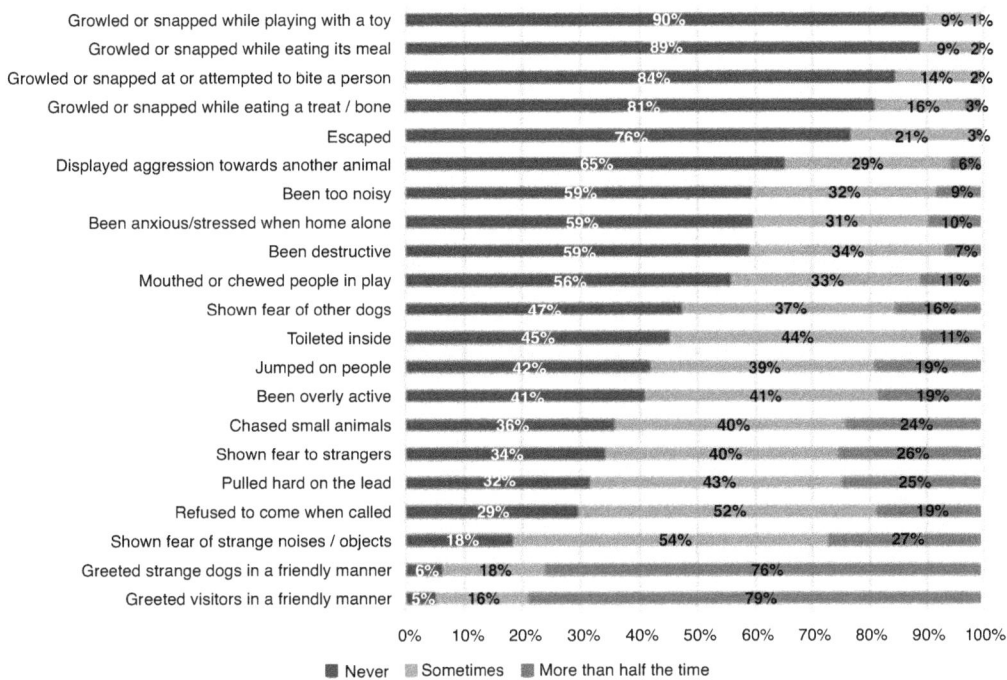

Fig. 5.3. Behaviour of imported rescue dogs (Norman *et al.*, 2020). Image is open access.

> The desire to remove free-ranging dogs from their natural living environment and place them in human homes as companions is widespread. This practice involves planning a new living condition for these animals, which is believed to be healthier than their free-ranging lifestyle. These new homes may be far from the areas where the dogs originally lived, possibly in another part of the world. Removing dogs from their natural environment and relocating them to unfamiliar settings disrupts their social structures and natural behaviours. This intervention can lead to unintended outcomes, including stress, trauma, challenges in adaptation, and other severe consequences, including euthanasia.
>
> (Adda, 2024)

Not only is the permanent location of the dog in many circumstances a concern, but the method is also problematic for many individuals. It is not uncommon to see imported dogs fearful of the car and other methods of transport, presumably because they have already been transported hundreds (sometimes thousands, or tens of thousands) of miles. This brings caregivers and professionals to another important treatment consideration, treating the fear or phobia of travel.

A key thing for humans to understand relates to the ethology of the free-ranging dog and the implications of such when thrust into a pet dog household. Many street dogs have spent their life employing survival strategies, managing their own relationships and environment and having total freedom over how to spend their time. This completely changes when a dog joins a human family, particularly if that family has unrealistic expectations of their new friend.

Puppy Farms

Intensive breeding is a problem for dogs all over the world. Before the introduction of Lucy's Law in 2020 in England (The Animal Welfare (Licensing of Activities Involving Animals)

Dog Type	Behaviour	Never (%)	Rarely (%)	Sometimes (%)	Often (%)	Always (%)	Do Not Know (%)	p-Value
	Owner home							
RD	whines for attention	29	30	30	9.9	1.5	0.3	<0.0001
FS	whines for attention	39	31	21	7.5	1.7	0.2	<0.0001
RD	follows owner	1.5	9.9	27	41	20	0.6	<0.0001
FS	follows owner	3.7	14	33	36	14	0.2	<0.0001
	Fear of humans							
RD	of women	86	9.2	3.5	0.9	0.2	0.4	<0.0001
FS	of women	71	18	7	13	1.1	0.5	<0.0001
RD	of men	71	16	9.2	3.0	0.6	0.6	<0.0001
FS	of men	45	19	21	12	2.9	0.5	<0.0001
RD	of children	68	15	10	3.4	1.9	1.1	<0.0001
FS	of children	55	20	12	7.8	3.0	3.0	<0.0001
	Fear of dogs, objects and sounds							
RD	unknown dogs	37	33	20	6.6	2.4	1.1	<0.0001
FS	unknown dogs	27	31	28	9.2	3.6	1.7	<0.0001
RD	unknown objects	25	44	24	5.8	1.0	0.8	<0.0001
FS	unknown objects	17	36	31	12	3.0	1.2	<0.0001
RD	of sounds	34	32	18	11	5	0.9	<0.0001
FS	of sounds	19	27	23	17	11	1.7	<0.0001
	Aggression							
RD	towards men	84	8.5	3.8	1.5	0.8	1.2	0.01
FS	towards men	79	9.0	5.4	4.5	1.2	1.1	0.01
Stress								
RD	on walks	55	28	12	3.0	1.2	1.1	0.04
FS	on walks	50	27	18	4.2	0.8	0.9	0.04
RD	unknown people	56	26	11	4.7	1.2	1.0	<0.0001
FS	Unknown people	41	29	16	8.4	5.1	0.8	<0.0001

Fig. 5.4. Responses to 11 questions from both owners of former street dogs (FS), and dogs reared in Denmark (RD) (Munkeboe et al., 2021). Shared with permission.

(England) (Amendment) Regulations, 2019) there was little in place to protect dogs from being farmed. Unfortunately, this illicit trade is still rife in parts of the UK as well as in other countries around the world, with enforcement of Lucy's Law proving almost impossible.

The common practice of buying and selling dogs online facilitates nefarious practices; a person only has to search for a puppy of their desired breed on Google and they can often collect their dog that same day. It has become increasingly common for puppy farmers to deceive well-meaning potential owners by inviting them to rented locations and giving excuses for the absence of the mother. Ross *et al.* (2023) collected data on 17,389 online puppy advertisements over a 2-year period in the UK, demonstrating the lucrative online trade in puppies. Wales is proven to be the 'most prolific region for sales of puppies, when calculated per-capita', motivated no doubt by a 2003 initiative for struggling farmers to supplement their income by dog breeding. Indeed, Wales is considered to be a prime location for puppy farms, though exact numbers are impossible to know. Perhaps the most worrying finding from this study was the prevalence of breeds with conformational disorders (CD) such as French Bulldogs and Chihuahuas in online advertisements (also seen in a study conducted by Paul *et al.*, 2022). Though CD breeds comprised only 3.4% of the 559 advertised breeds, they were advertised significantly more than others (totalling 46.8% of advertisements). Of course, not all of these dogs will have originated from puppy mills, but a lack of regulation makes it very easy for criminal sellers to sell quietly alongside ethical, regulated breeders.

In the USA, 'commercial breeding establishments' (CBEs) still supply some pet stores and websites. Notwithstanding the poor conditions that these animals are frequently exposed to, their emotional wellbeing is often further impacted by behaviour issues caused by trauma.

The RSPCA Australia Knowledge Base defines a 'puppy farm' as 'an intensive dog breeding facility, operating under inadequate conditions' (RSPCA Australia, n.d.). Welfare concerns raised by the RSPCA are as follows:

- Lack of basic essentials.
- Lack of space.
- Lack of safety.
- Lack of adequate housing.
- Lack of care (grooming, preventative care, veterinary care).
- Lack of socialization.
- Lack of genetic planning (not selecting for health or temperament).
- Lasting trauma.

Behaviour problems are common in farmed dogs. Dogs obtained from puppy mills (puppy farms or CBEs) have been shown in some studies to score higher than those from independent breeders in aggressive or unfriendly behaviours as well as fearfulness (Bennett and Rohlf, 2007; Casey *et al.*, 2014), with some studies reporting dogs from CBEs up to 60% more likely to display aggressive behaviour (McMillan *et al.*, 2011). There are many possible contributors to these dogs' reduced ability to cope. The compromised health and wellbeing of the mother (McMillan *et al.*, 2011), barren environment and lack of selection for temperament in breeding adults are thought to be just a few factors leading to trauma. The importance of socialization is commonly spoken about with regards to puppies, and this process starts significantly before they are handed over to their new family. However, in the sterile environment of a puppy farm a lack of exposure to novel stimuli is devastating for their development.

As well as traumatized puppies, breeding dams and sires are often rehomed once they can no longer produce puppies. Well-meaning caregivers who take these dogs home should also expect them to need significant ongoing support. McMillan *et al.* (2011) found that owners and fosterers of rehomed breeding dogs were significantly more likely to report both health and behavioural problems than their traditionally bred counterparts. A 2024 interview-based study supported these findings, with owners reporting 'worn out' dogs, some of whom were rehomed with prolapses, holes in their ears and even evidence of botched caesareans (Walliss, 2024). As seen in farmed puppies, breeding dogs have also been observed to display more behaviour problems, particularly those motivated by fear.

Caregivers that have fallen prey to puppy farmers and criminal networks should expect to take a trauma-informed approach to their puppy and invest time in extensive behaviour modification. This is also the case for caregivers of rehomed breeding dogs, the likes of which will typically bear both behavioural and physical scars of their former life. As with the aforementioned overseas rescue dogs, puppy mill victims will need time and understanding in order to cope with their new lives.

Dogs Impacted by Criminal Activity

Initially, this section should be prefaced by saying that there are many dogs in the UK sadly impacted by the Dangerous Dogs Act (1991) (DDA) who have lived normal lives without experiencing trauma. It is by no means implied that dogs such as American Bully XLs and Pit Bull Terriers are *automatically* associated with criminal activity.

However, those dogs that have been exposed to criminal activity (including dog fighting) are undoubtedly trauma survivors and require TIC. Often the dogs chosen for this sort of lifestyle are those that look intimidating and so bull breeds and mastiffs are common targets.

Despite dog fighting being illegal in the UK since the 1800s, between the years of 2006 and 2015 the RSPCA received 4855 complaints regarding organized dog fighting (BBC News, 2017). Even if dogs are not directly exposed to fighting, there is still the possibility that they will have been used as status dogs and exposed to violence amongst humans.

Although as previously mentioned, many dogs impacted by the DDA will have lived conflict-free lives, it would be an oversight not to consider the amount of trauma that American XL Bullies have been exposed to since being added to the DDA in 2024. The implications of this change to the DDA are likely to have impacted more than 220,000 dogs across the UK (PDSA, 2024). Following their addition to the Act, they have joined the demographic of dogs that are legally required to wear a muzzle and be on-lead at all times when in public.

A drastic lifestyle change, such as that caused by the amendments to the DDA, could easily be traumatic for any dog. Dogs that previously enjoyed ample freedom have had to adjust to a new lifestyle led by restrictions. As demonstrated by the overseas rescue dogs discussed above, experiencing a sudden lack of agency can cause a drastic change in worldview leading to many fear- and frustration-based behaviours. Caregivers may have noticed an increase in impulsivity leading to reactive behaviour, or they may feel that their dog has become shut down and depressed.

Supporting a Traumatized Dog (Using the 'Three Pillars of Trauma-Informed Care')

Once a traumatized dog has been identified (either by history taking, or through behavioural presentation) a trauma-informed treatment plan should be using the 'three pillars of trauma-informed care'. Caregivers should understand that a primary adaptation phase will take significant time (Thumpkin *et al.*, 2024), typically at least 6 months (Salgirli Demirbas *et al.*, 2014).

The Thumpkin *et al.* (2024) study also emphasizes the caregiver's need to adjust their expectations, be patient and create a safe environment for the dog as well as the household. Taking the time to build trust and learn to respect one another's boundaries are also important.

Pillar 1: Safety

As per the first pillar of TIC, creating *safety* should be a priority. This encompasses safety in the environment, in relationships and in access to resources.

Rachel Leather's Trauma-Informed Treatment Model

A fundamental treatment strategy for traumatized dogs is to follow behaviourist Rachel Leather's 'Trauma-Informed Treatment Model' (Leather, 2020).

Firstly, a 'Needs Assessment' is performed based on biological, psychological and social needs. Strategies are then formed to meet any unmet needs and all needs are assigned a 'RAG' (red, amber or green) status based on priority. An example would be checking that a dog has recently been assessed by a veterinarian, assigning 'red' status (high priority) if the answer is 'no'. This enables caregivers to ensure that their dog is healthy and on a consistent emotional baseline without any unmet needs before starting their Support Plan.

Afterwards, caregivers complete the Support Plan, organized by RAG status. Caregivers are provided with measurable actions in order to meet the needs of the dog and are asked to tick them off as they work through them. Primary considerations for each step of the model are as follows:

Step 1 'I feel safe somewhere': The dog has shelter and a safe space (see below) with their fundamental needs met.
Step 2 'I feel safe with someone': The dog can engage with at least one individual in the home, is able to give eye contact and be touched.
Step 3 'I feel safe in my environment': The dog is not immediately triggered by anything in the home or regular visitors to the home. This extends to the front and back garden.
Step 4 'I feel safe exploring the outside world': The dog is able to go to a variety of places, travel in the car and deal with some element of novelty.
Step 5 'I feel safe': The dog is able to cope beyond the predictable, investigate new things and problem solve without panicking.

Using this plan ensures that a dog's triggers are not overlooked and that their needs are consistently met, creating an overall sense of safety and eventually changing their worldview. Fig. 5.5 demonstrates how the treatment plan might be used.

Safe haven

A 'safe haven' is a fundamental principle for all animals. It is even more pertinent when an animal is traumatized and needs a place both to hide but also to recover and decompress.

Typically, a safe haven is a place such as a crate, bed or cupboard under the stairs. However, it is important to remember that a chosen human can also represent a safe space when a dog is out and about (Mariti *et al.*, 2013: Payne *et al.*, 2015). Normally, a dog will start to gravitate towards a certain place that can then be adapted into their haven of choice.

For the purposes of this exercise a crate will be the recommended place for the establishment of a safe haven. The key principles are:

- The safe haven is accessible as often as possible, or the dog should easily be able to ask for access.
- The safe haven is not encroached upon by humans or other animals, avoid coaxing the dog out at all costs.
- Nothing bad happens to the dog when they are in the safe haven (for example, it should never be used as punishment).

Step One: 'I feel safe somewhere'

Criteria	RAG Status	Steps to meet criteria	Criteria met
Dog has a 'safe haven' freely available	Red	Provide bed in location that is freely accessible but cannot be used by other animals. Reinforce for using it.	No
Dog has access to fresh food and water	Red		Yes
Dog is not under immediate threat from other animals in the home	Red	Separate new dog from established dog with a stair gate until safe to start gradual introductions.	No
Dog has a safe place to toilet	Red		Yes
Dog's immediate health needs are met	Red		Yes

Fig. 5.5. Starting with Step One 'I feel safe somewhere', criteria are given that will help a dog to feel a sense of safety somewhere in the home. Criteria are graded by 'RAG' status, then guidance is provided on how to meet any unmet criteria. Image is author's own, using the ideas of Rachel Leather (Leather, 2020).

- A safe haven typically takes time to build and should not be rushed.
- The space should be comfortable; many dogs like to have it covered.
- If there is a part of the home that your dog prefers when they feel frightened you should put the safe haven here.

Your dog may use the safe haven organically. Or, you may have to encourage them for a short while using treats or other reinforcers. Once they start to use the space voluntarily, they are likely to choose it when they feel worried.

Resource safety

Dogs who have learned survival strategies and self-preservation may demonstrate defensive behaviours around resources (often known as 'resource guarding'). It is important for caregivers to understand why this is happening, but additionally:

- Provide an abundance of resources and ensure they are accessible.
- Consider any barriers that may make resources for unattainable for a frightened dog.
- Do not remove things from the dog by force unless it is an immediate threat to their safety.
- Leave the dog to eat undisturbed (including away from other pets).
- Stop providing a resource if it consistently creates conflict in the home.

Consent-based handling

Many aspects of handling are very worrying for traumatized dogs. This can be in relation to the dispensing of medication, bathing or coat care. Or a dog might be frightened of equipment such as the collar or harness. They may also be suspicious of affectionate handling and choose to keep away from humans when they are initially introduced to the home. Principles for avoiding handling stress include:

- Prioritize what is absolutely necessary; baths and groomer visits are incredibly stressful and should not be done without ample desensitization and preparation beforehand.
- Adopt a Co-Operative Care protocol (see Chapter 9).
- Desensitize the dog to equipment over time; it may be sensible to start temporarily with a slip lead on the fatty part of the dog's neck as the piece of equipment requiring the least human contact.

It is also highly advised to use the '3 Second Rule' when petting or handling, as well as passing this onto other family members and visitors. Simply put, this rule advises the individual to pet the dog for 3 seconds before pausing to see if the dog wants the interaction to continue. If they want to continue, they will likely move closer, lick or nudge the handler. If they would rather not continue the interaction, they will likely move away. If the dog is unclear in their decision, it is best to leave them alone as a precaution.

Manage exposure to triggers

It is difficult for a dog to feel safe when they are consistently exposed to stressors. These things could be typical household appliances or activities that seem mundane to us, but a traumatized dog could find them incredibly challenging. Basic things a caregiver can do to manage exposure to triggers include:

- Covering windows.
- Consistent background noise.
- Avoiding or reducing walks until the dog is ready.
- Encouraging the dog to use their safe space when 'scary' things are happening (for example, hoovering).
- Prioritizing activities (for example, avoiding a bath as discussed above).

Caregivers should also be very mindful of arousal and trigger stacking, as discussed in Chapter 6.

Pillar 2: Connections

Building secure relationships

A secure attachment style can be described as a 'healthy attachment style that enables individuals to work autonomously as well as with others when appropriate' (Little *et al.*, 2009). Taking into account the fact that many traumatized dogs will have learned to protect themselves and work alone, they may struggle to work alongside humans and other animals. So, caregivers should seek to build secure relationships as a priority by doing the following:

- Being predictable and consistent in their interactions.
- Not leaving the dog alone for longer than they can cope with.

- Being a symbol of good things and avoiding negative associations (this is discussed more below).
- Starting basic positive reinforcement training.
- Being responsive to what the dog is communicating.

Employing kind and ethical training methods

As demonstrated in Chapter 6, 'Emotional Wellbeing While Learning', an easy way to damage the relationship between dog and caregiver is to use punitive or 'balanced' training methods. Traumatiszd dogs with a tendency to be suspicious and pessimistic will really struggle to trust any human that is heavy handed towards them. While this may be a preferable 'quick fix' for some individuals, the consequences are likely to become apparent later on.

Pillar 3: Managing emotions

When using TIC to treat traumatized children, a fundamental component is teaching them to manage their emotions in situations that they find difficult. The same principle can be applied to dogs.

Managing arousal

Taking steps to limit over-arousal in a traumatized dog will drastically increase the likeliness of stressful incidents. Consider 'trigger stacking' (refer to Chapter 6) when planning their day and allow ample time for decompression and recovery. Ideally, dogs should be slowly desensitized to triggers over time before being introduced in order to keep arousal down.

Confidence building

Traumatized dogs will benefit from confidence building to prepare them for exposure to novelty. Allowing them to explore at their own speed, dogs should be introduced to a variety of textures, objects and noises. Choose an open but secure and familiar space and provide an assortment of stimuli. Dogs can be encouraged to investigate by hiding things like toys and high-value food rewards amongst the stimuli, but they should not be coerced or forced.

Some items may be difficult to start with, but over a period of time their confidence and exploratory nature will develop. By giving them the tools to cope with novelty, dogs will be less likely to panic when exposed to new things.

Case vignette: Jerry

Fig. 5.6. After being liberated from the slaughterhouse in China, Jerry was emaciated and had a fungal skin infection. Image is author's own.

Fig. 5.7. Once settled in his new home using TIC, Jerry put on weight and his coat healed. He also had extensive dental surgery to relieve pain from his mouth. Image is author's own.

Jerry is a 4-year-old Shiba Inu. His caregivers initially became aware of him when he was rescued from a slaughterhouse in China, subsequently being transferred to a well-meaning but poorly equipped rescue centre where he lived for an extended period in a small, wire crate (Fig. 5.6).

When Jerry was first brought over to the UK, he was shut down and did not know how to behave like a pet dog. Having been rehomed to London, Jerry constantly experienced stressors (even if in his aloof Shiba type way he did not show it!). His caregivers found it difficult to reliably encourage him to eat, he did not know how to play, and he did not really show any affection towards them. He also reacted to other dogs while on walks, a behaviour he was not thought to have displayed at the group centre in China.

Jerry was treated using Trauma Informed Care, first prioritizing his welfare needs and removing any additional pressure. Often this meant choosing not to walk Jerry, understanding if he did not want to eat, and not expecting any physical affection from him. In order to support him, his caregivers made sure that he was treated for any pain (leading to various check-ups and eventually some dental surgery), he was made to feel safe in the home and never made to feel frightened, and he was provided with a routine to help bring predictability to his life.

Over time Jerry began to slowly come out of his shell and show his cheeky personality! He chose some squeaky toys and started to engage in play with his caregivers. He started to come up onto the sofa for cuddles, otherwise he would bask on his back in the sun with his legs in the air. He even started to make friends with other dogs. By significantly reducing expectations and helping Jerry to adjust in his own time, Jerry has flourished into a funny and adventurous little dog (Fig. 5.7).

Real-Life Application

Traumatized dogs require a careful approach akin to the principles laid out in TIC for children. The 3 pillars to focus on are 'Safety, Connections and Managing Emotion'. In doing these things, caregivers and professionals can help traumatized dogs to adjust to their new surroundings and eventually adapt to their new lives.

This is likely to take time and patience and is often not without setbacks. Therefore, caregivers and practitioners alike must go into these relationships informed and ready to work hard. With this in mind, helping to rebuild and repair the emotional wellbeing of a traumatized dog, teaching them to trust and enjoy life once again, is a deeply rewarding experience.

References

Adda, M. (2024) Wellness or hellness? Rediscussing free-ranging dogs welfare. *Multispecies Communities and Narratives* 5–43.

Bath, H. (2008) The three pillars of trauma-informed care. *Reclaiming Children and Youth* 17(3), 17–21.

BBC News (2017) Dog fights prompt 5,000 calls to RSPCA in past decade. Available at: https://www.bbc.co.uk/news/uk-england-38653726 (accessed 11 July 2025).

Bennett, P.C. and Rohlf, V.I. (2007) Owner-companion dog interactions: Relationships between demographic variables, potentially problematic behaviours, training engagement and shared activities. *Applied Animal Behaviour Science* 102(1–2), 65–84. DOI: 10.1016/j.applanim.2006.03.009.

Casey, R.A., Loftus, B., Bolster, C., Richards, G.J. and Blackwell, E.J. (2014) Human directed aggression in domestic dogs (*Canis familiaris*): Occurrence in different contexts and risk factors. *Applied Animal Behaviour Science* 152, 52–63. DOI: 10.1016/j.applanim.2013.12.003.

Collins, K., Miller, K., Zverina, L., Patterson-Kane, E., Cussen, V. *et al*. (2022) Behavioral rehabilitation of extremely fearful dogs: Report on the efficacy of a treatment protocol. *Applied Animal Behaviour Science* 254, 105689. DOI: 10.1016/j.applanim.2022.105689.

Corridan, C.L., Dawson, S.E. and Mullan, S. (2024) Potential benefits of a 'trauma-informed care' approach to improve the assessment and management of dogs presented with anxiety disorders. *Animals* 14(3), 459. DOI: 10.3390/ani14030459.

Dangerous Dogs Act (1991). Available at: https://www.legislation.gov.uk/ukpga/1991/65/contents (accessed 11 July 2025).

Leather, R. (2020) *Recovery from trauma: A route-map for assessment and treatment planning*. Control the Meerkat Conference. Available at: https://conference.controlthemeerkat.com/wp-content/uploads/2021/07/Psychological-trauma-in-dogs-A-route-map-for-assessment-and-treatment-planning-1.pdf (accessed 6 March 2025).

Little, L.M., Nelson, D.L., Gooty, J. and Simmons, B.L. (2009) Secure attachment: Implications for hope, trust, burnout, and performance. *Journal of Organizational Behavior* 30, 233–247.

Mariti, C., Ricci, E., Zilocchi, M. and Gazzano, A. and (2013) Owners as a secure base for their dogs. *Behaviour* 150(11), 1275–1294.

McMillan, F.D., Duffy, D.L. and Serpell, J.A. (2011) Mental health of dogs formerly used as "breeding stock" in commercial breeding establishments. *Applied Animal Behaviour Science* 135(1–2), 86–94. DOI: 10.1016/j.applanim.2011.09.006.

Munkeboe, N., Lohse-lind, A., Sandøe, P., Forkman, B. and Nielsen, S.S. (2021) Comparing behavioural problems in imported street dogs and domestically reared danish dogs—the views of dog owners and veterinarians. *Animals* 11(5), 1436. DOI: 10.3390/ani11051436.

Norman, C., Stavisky, J. and Westgarth, C. (2020) Importing rescue dogs into the UK: Reasons, methods and welfare considerations; Importing rescue dogs into the UK: Reasons, methods and welfare considerations. *Veterinary Record* 186(8), 248–248. DOI: 10.1136/vetrec-2019-105380.

Paul, E.S., Coombe, E.R. and Neville, V. (2022) Online dog sale advertisements indicate popularity of welfare-compromised breeds. *Journal of Applied Animal Welfare Science* 27(4), 702–711. DOI: 10.1080/10888705.2022.2147008.

Payne, E., Bennett, P.C. and McGreevy, P.D. (2015) Current perspectives on attachment and bonding in the dog–human dyad. *Psychology Research and Behavior Management* 8, 71–79. DOI: 10.2147/PRBM.S74972.

PDSA (2024) *PDSA Animal Wellbeing (PAW) Report*. Available at: https://www.pdsa.org.uk/what-we-do/pdsa-animal-wellbeing-report/paw-report-2024 (accessed 10 July 2025).

Ross, K.E., Langford, F., Pearce, D. and McMillan, K.M. (2023) What patterns in online classified puppy advertisements can tell us about the current UK puppy trade. *Animals* 13(10), 1682. DOI: 10.3390/ani13101682.

RSPCA Australia (n.d.) What is a puppy farm? RSPCA Australia Knowledge Base. Available at: https://kb.rspca.org.au/knowledge-base/what-is-a-puppy-farm/ (accessed 5 March 2025).

Salgirli Demirbas, Y., Emre, B. and Kockaya, M. (2014) Integration ability of urban free-ranging dogs into adoptive families' environment. *Journal of Veterinary Behavior* 9(5), 222–227. DOI: 10.1016/j.jveb.2014.04.006.

The Animal Welfare (Licensing of Activities Involving Animals) (England) (Amendment) Regulations 2019 (2019) Pub. L. No.1093 (n.d.). Available at: https://www.legislation.gov.uk/uksi/2019/1093/contents/made (accessed 11 July 2025).

Thumpkin, E., Pachana, N.A. and Paterson, M.B.A. (2024) Coming home, staying home: Adopters' stories about transitioning their new dog into their home and family. *Animals* 14(5), 732. DOI: 10.3390/ani14050723.

van der Kolk, B. (2014) *The Body Keeps The Score*. Penguin Books, London.

Walliss, J. (2024) "You don't have the right to steal my life": Exploring the harms of puppy farming on ex-breeding dogs. *Journal for Human-Animal Studies* 10, 38–61.

Zeiman, V.M. (2015) Adopting a former street dog - a comparison of street and pet dog owner personality, dog characteristics and relationship. MSc Thesis, University of Vienna, Austria.

6 Emotional Wellbeing While Learning

Jade Nicholas*

CAB, Winchester UK

Abstract

This chapter considers a dog's emotional wellbeing when they are in a learning environment. Animals are consistently learning, but pet dogs are frequently exposed to controlled learning in the form of training and behaviour modification. Reward-based methods (primarily positive reinforcement with some negative punishment) have been widely proven to help dogs learn more efficiently, increase optimism and enhance a bond with their handler. This chapter acknowledges, however, that many individuals still rely heavily on positive punishment or 'balanced' methods and so addresses these methods' shortcomings as well as introducing The LIFE Method as a model for all handlers to follow. Other concerns regarding controlled learning are explored, including the efficacy of equipment, a handler's choice of environment and the optimum level of emotional arousal for success in training and behaviour modification.

Introduction

> Twenty to 30 years ago, social dominance theory and ideas about wolf behavior in the wild guided dog-training methods, which focused on punishing bad behaviors by using choke chains, pinch collars, and electronic collars because wolves in the wild appeared to gain higher rank through force.
>
> (Yin, 2007)

To fit into our complex human society, dogs are expected to adapt to various environments and routines. Because of this, something held in high regard is the extent and reliability of a dog's training, commonly referred to as 'obedience'. For example, an 'obedient' dog is considered to be responsive (without fail) to their human, walk beautifully on the lead, recall perfectly and generally behave without concern. The problem is that a consistent power struggle caused by the expectation of obedience can rapidly lead to diminished welfare. An extreme example of this is shown in Dr Sophia Yin's 2007 paper *Dominance Versus Leadership in Dog Training* in which she describes being advised by a professional to 'alpha roll' her dog before hitting it on the nose, followed eventually by choking him to the point of passing out. Dr Yin was told in this instance that 'it [was] important to win all aggressive encounters'.

It should not be difficult to understand why the above-described methods are detrimental to a dog's welfare; however, perceived milder punishment methods are more widely accepted in society. The 2024 PDSA Paw Report found that 22% of owners had used at

---

*Corresponding author: aboutyourdog@outlook.com

least one form of aversive training on their dog, with some respondents using multiple (PDSA, 2024).

All animal behaviour is caused by a combination of genetics, environmental factors, health and prior learning (to name just a few contributors), and so to see a dog as either 'obedient' or 'disobedient' is hugely problematic. Collins Dictionary (n.d.) defines obedience as 'the act or an instance of obeying; dutiful or submissive behaviour'. Given the multifaceted reasons a dog might perform a behaviour, it is therefore negligent at best to insist on submission without opportunity for individual expression. At worst, it is nothing but cruel.

For example, a dog may be considered 'disobedient' if he reacts to another dog against the will of his handler, but the complex field of animal behaviour tells us that a reaction can happen for reasons such as pain, fear or frustration (amongst other things). For this reason, the ethical consequences of prioritizing obedience over an investigation into the emotional motivation of that dog are many.

Humans value order and obedience highly among their own species as well as others, so it can be challenging for a more ethical school of thought to take precedence. Studies in human psychology demonstrate the importance of obedience in society, stating that 'for many people, obedience is a deeply ingrained behavior tendency, indeed a potent impulse overriding training in ethics, sympathy, and moral conduct' (Milgram, 1974). This importance is even more apparent when looking at obedience as a measure of value in dogs, with Tunaitytė *et al.* (2024) having found that perceived 'obedience' is a positive predictor for dogs being adopted from shelters.

This chapter explores the consequences of training methods on an individual dog including the successful learning and retention of knowledge, but also the long-term consequences for a dog's emotional wellbeing. It seeks to promote a fun and empowering way for dogs to learn without having to resort to any harmful practices in the name of 'obedience'.

How Dogs Learn

Like all animals, dogs learn through a series of experiences, associations and consequences. As well as this, they are known to learn through observations and social interactions with both their own species and others (Pongrácz, 2014). Essentially, dogs learn in the same way as humans and most other animals.

There are some fundamental theories underpinning the learning experience of mammals, and these are explored in greater detail below. Further reading is recommended for those unfamiliar with learning theory, however, as this will only scratch the surface.

Classical conditioning

Classical conditioning is a learning theory initially made famous by Russian physiologist Ivan Pavlov in the 1800s (Rehman *et al.*, 2017) (Fig. 6.1).

Pavlov theorized that a previously unconditioned stimulus (US) (a stimulus with no meaning or association) could be paired with a conditioned stimulus (CS) (a stimulus with meaning or emotional attachment). The US would start to predict the presence of the CS

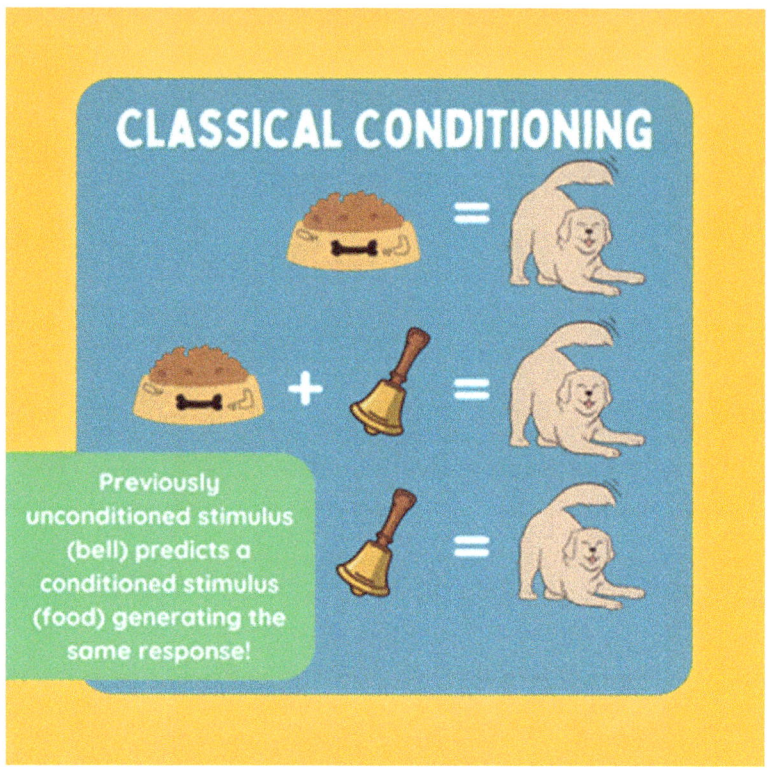

Fig. 6.1. Classical conditioning. Image is author's own.

in turn generating the same response. This is often referred to colloquially as 'learning by association'.

In the famous experiment now known as 'Pavlov's dogs', Pavlov noticed that the presence of food would cause the dogs to drool. He first introduced the bell as a predictor of food arriving before removing the food and using the bell on its own. As predicted, the dogs started to salivate upon being presented with the bell, demonstrating that they had formed a learned association with the formerly unimportant object.

An easy example of this technique used in modern day application is when teaching a dog to recall. By taking a whistle (US) and a high-value food reward (CS), classical conditioning can be applied in this way:

1. Blow the whistle and drop a piece of food on the floor at your feet.
2. Repeat over and over for a series of days.
3. Start to increase the distance away from your dog and watch them come back to your feet when they hear the whistle.

Whistle = Food = Go to human

As training goes on dogs are able to respond to a more sophisticated version of the protocol (Rehman *et al.*, 2017). For example, some scenarios may require a dog to recall to a specific whistle but to offer another behaviour for a second whistle.

Operant conditioning

The second fundamental learning theory is operant conditioning, also known as 'learning by consequence or 'the four quadrants'. Burrhus Frederic Skinner established operant conditioning in 1948 using a controversial experiment now known as 'Skinner's Rats'. Rats were placed in an experimental box with a food dispenser and an electric floor connected to a lever. Through observing various behaviours performed by the rats, Skinner designed the four quadrants. Skinner established three responses that can follow a behaviour:

1. **Neutral:** The probability of a behaviour being repeated is neither increased nor decreased.
2. **Reinforcer:** Probability of a behaviour being repeated is increased.
3. **Punisher:** Probability of a behaviour being repeated is decreased.

The dog's appraisal of the consequences following their behaviour (neutral, positive or negative) will then determine what happens next.

As seen in Fig. 6.2, these findings have established four quadrants of operant conditioning which are used in modern dog training. Here are some common examples of their application as shown in the figure:

- **Negative Reinforcement:** The use of a slip lead to create pressure until the dog stops pulling and releasing pressure when the dog walks on a loose lead, increasing loose-lead walking.

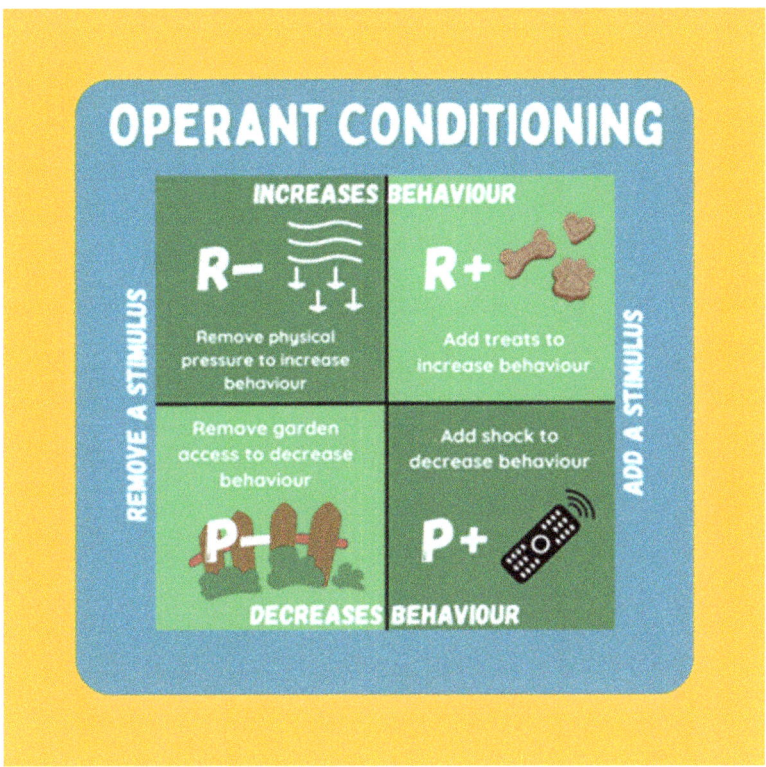

Fig. 6.2. Operant conditioning. Image is author's own.

- **Negative Punishment:** Removing access to a garden in order to decrease barking behaviour.
- **Positive Reinforcement:** Giving treats to acknowledge a desirable behaviour in order to increase it.
- **Positive Punishment:** Administering a shock via a shock collar when a dog reacts to a trigger, therefore reducing the likelihood of repetition.

Commonly Used Training Methods

> Currently, although the Animal Behaviour and Training Council (ABTC) sets and maintains the standards of knowledge and practical skills needed to be an animal trainer, training instructor or animal behaviour therapist, there is no legal requirement for pet trainers or behaviourists to have any qualifications at all, and it can therefore be extremely difficult for an owner to know whose advice to follow. Using outdated or inhumane training techniques can have a detrimental effect on a dog's welfare and behaviour, so further regulation and guidance are needed.
>
> (PDSA, 2023)

There is much debate about the correct way to train dogs. One thing that is not debated, however, is the need for appropriate training in the prevention of behavioural problems. When dogs do not receive the training and behavioural support that they need they are at risk of relinquishment or in some cases euthanasia. Training also helps caregivers to keep their dogs safe in public and comply with animal laws.

As expressed by Dr Sophia Yin in her 2007 article, less than a century ago the most popular methods of dog training involved painful punishments such as electric shock collars and physical corrections. The infamous 'Koehler Method' established in 1946 is one of the more brutal methodologies to have been used for dog training. Koehler and many other practitioners of the time placed a lot of value on establishing 'dominance' and recreating dynamics seen in packs of wolves (Lenehan, 1986), mentioned previously in Chapter 3.

Conversely, modern training has moved away from punitive methods and relies mostly on the positive reinforcement quadrant. Some ethical methods require the use of the negative punishment quadrant, but in doing so every effort is made not to compromise the welfare of the dog. For example, a modern trainer might cover a window with newspaper to block the view of a trigger (negative punishment), but will be sure to reinforce the dog coming away from the window and redirect them onto an alternative, positive activity (positive reinforcement).

Some individuals practice using 'balanced' methods, a combination of positive punishment and positive reinforcement, though often the ratio is not 50:50. As of 2018, only around 16–20% of the UK population could confidently say that they only used reward-based (positive reinforcement and negative punishment) training methods (Todd, 2018).

All four quadrants are valid and yield results. However, as a professional or caregiver the ethics of each quadrant should be at the forefront of our considerations. From an emotional wellbeing perspective, the ethical benefits of positive reinforcement are in stark contrast to concerns raised by the use of positive punishment.

Positive reinforcement versus positive punishment

Examples of positive reinforcement methods include:

- Clicker training.
- Using food, toys or praise to reward.
- Environmental management to prevent undesirable behaviours.
- Training through play.
- Co-operative care.

Examples of positive punishment methods include:

- Electronic shock collars.
- Prong collars.
- Slip leads in garotte position used in an upward correction.
- Spray cans or water pistols.
- Physical punishment (slapping, hitting).

Referring once again to the Koehler method, an 'extreme' example of positive punishment is shared in this 1986 magazine article:

> To control an aggressive dog in extreme circumstances—in other words, to stop a fighter or biter—they sometimes "hang" the dog by its collar, literally letting it twist in the wind. They say that mother dogs discipline their puppies in a similar manner, taking them by the scruff of the neck and holding them, sometimes shaking them, in midair; Dick Koehler says the method is effective because dogs fear the "loss of environmental control" that they suffer when their feet lose contact with the ground.
>
> (Lenehan, 1986)

Many would hope that the method of hanging a dog in order to 'train' them is outdated or even better, illegal. However, a lack of regulation of dog trainers and behaviourists in the UK sees unqualified 'professionals' and rookie 'owners' utilize these methods in the name of obedience even as this book is published in 2025.

In 2013, a case study was published by Grohmann *et al.* detailing the consequences of these methods when applied to a 1-year-old German Shepherd (Grohmann *et al.*, 2013). The dog in question was disciplined using a similar hanging method to the one made famous by Koehler, suspended off the ground by his choke chain collar. Initially, the dog started circling before losing consciousness and ultimately, veterinary scans showed multiple forms of brain damage. Tragically this dog was euthanized due to his symptoms and the irreversible damage caused.

One might suggest that the commonly used punitive methods of the 21st century are significantly less damaging than those brutally imposed 80 years ago. However, there are still measured consequences of using positive punishment in dog training, regardless of how harmless they profess to be. For example, one known impact of training method choice is the relationship formed with a dog's handler. One study (Vieira de Castro *et al.*, 2019) measured a dog's response to their handler versus a stranger in what is known as a 'Stranger Situation Test'. This study found that dogs trained using positive reinforcement methods played more in the presence of their owner than the stranger, and that they were more likely to follow and enthusiastically greet their owner. Studies such as this one performed by Vieira de Castro *et al.* (2019) would suggest that the emotional state of a dog in the presence of a caregiver using positive reinforcement methods is more positive than those utilizing positive punishment.

Not only is there evidence to suggest that the use of punishment causes the dog to interact differently with their handler, studies suggest that a significant portion of dogs (particularly those presenting with human directed aggression) display hostility towards their handler when aversive methods are used (Herron *et al.*, 2009). Conversely, the 2009 study found that positive reinforcement 'elicited aggression in very few dogs, regardless of presenting complaint'.

There has never been any scientifically proven benefit to using punishment-based methods over rewards (Ziv, 2017); in fact, punishment is thought to be associated with a greater number of presenting behavioural problems (Blackwell *et al.*, 2008), and so once again the choice of methods used raises an important ethical question. Why would we choose to hurt our dogs when we can effectively achieve the same result using rewards? Further still, why would we risk damaging our relationship when it is completely unnecessary to do so?

LIMA and LIFE Methods

Given a lack of cohesive approach to training and behaviour modification in the animal training and behaviour industry, there has been a pressing need for a model on which to base treatment choices. Initially, the 'Least Intrusive Minimally Aversive' (LIMA) model designed in 2005 was provided as a recommended framework and many science-led practitioners began working in accordance with this model.

However, LIMA generated scrutiny for being ambiguous; in particular, LIMA does not give recommendations for good practice. Further still, in this evolving welfare-led field practitioners have criticized LIMA for justifying the use of aversive tools. LIMA encourages practitioners to start with the lowest ranking method in terms of severity and advance to more severe tools should the behaviour not improve. The informed behaviourist or caregiver knows, however, that a lack of improvement in behaviour does not warrant a firmer hand, rather further investigation leading to a change of diagnosis. LIMA has been seen to give the 'green light' to aversive tools like prong and shock collars provided the handler has tried a lighter touch beforehand.

In 2024, the 'Least Inhibitive, Functionally Effective' (LIFE) model was published by Eduardo Fernandez (Fig. 6.3). This exciting new approach has now been adopted by reputable organizations such as the ABTC, the well-respected register for animal behaviourists in the UK. The vision on which the LIFE model was designed is as follows:

> A critical feature for effectively describing modern animal training and behavior change procedures should be a simple framework that connects theory, science, and practice. In other words, the framework should accurately detail both how to think about and do ethical animal training.
>
> (Fernandez, 2024)

Increasing meaningful choices

Focusing on the wellbeing of the animal, this part of the LIFE model focuses on maximizing choice. Fernandez (2024) accurately points out that a handler might profess only to

A Modern Approach To Thinking About Ethics and Animal Training

LEAST INHIBITIVE, FUNCTIONALLY EFFECTIVE

The **Least Inhibitive, Functionally Effective (LIFE)** approach provides a framework that adheres closely to the behavioural and welfare-focused sciences. It considers the impact of training methods on the wellbeing of both human and non-human lives.

The LIFE approach emphasizes the important interplay between **training success** and **positive welfare.**

INCREASE MEANINGFUL CHOICES

Least inhibitive means removing choice restrictions to increase quality of life.

- Avoid environmental restrictions to motivate behaviour, such as food deprivation.
- Improve meaningful options by expanding response alternatives and behavioural repertoires.

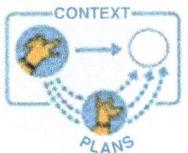

IDENTIFY BEHAVIOURAL FUNCTIONS

Assess causes of a behaviour so that we can directly connect them to behaviour change plans.

- Collect data based on observations and behaviour change manipulations.
- Match causes of a behaviour so that we can directly connect them to behaviour change plans.

MAXIMIZE TRAINING SUCCESS

Effectiveness is only one component of success. Impact of training on welfare is also important.

- Consider other welfare impacts, such as context and human–animal interactions.
- Training success includes positive welfare during and after training sessions.

Fig. 6.3. The LIFE model (Fernandez, 2024). Used with permission.

use the positive reinforcement quadrant, but they can still have limited choices available to their dog, which in turn compromises welfare. For example, your dog could be partaking in positive loose-lead walking training, but if they have no agency over where they are able to sniff or go to the toilet their welfare is compromised.

Further still, the handler should avoid manipulating the environment or access to resources in order to reach a behavioural goal. Otherwise, the handler risks coercing the animal into a behaviour they do not want to perform. For example, if you put a dog's food bowl next to another to encourage proximity you will likely succeed in bringing the animals closer together, but you will not diffuse any underlying conflict. Whilst environmental manipulation is a valid tool in behaviour modification, it should seek to compromise a dog's agency as little as possible.

Identify behavioural functions

Successfully identifying the function of a behaviour provides greater opportunity for meeting the needs of an animal. This part of the LIFE model encourages caregivers and practitioners to consider what motivates the dog to behave in a certain way and adapt accordingly. Choosing to ignore or suppress a behaviour, particularly when using positive punishment, can arguably disregard what the dog is trying to communicate.

A great example of failing to identify a behavioural function is that of a handler punishing (possibly with a slip lead, prong collar or e-collar) a dog for reacting out of fear to a stimulus. Not only is this completely wrong from the perspective of learning theory, it ignores the emotional state of the dog. When applying the LIFE model in this example, a practitioner would identify the dog's act of self-preservation from the perspective of fear and support them through ethical rehabilitation.

Maximize training success

The final primary component of the LIFE model focuses on the experience being 'functionally effective'. Fernandez asks the caregiver or practitioner to consider not only the immediate effectiveness of the training (based on the above understanding of behavioural function), but also the overall welfare experienced by the animal as well as their handler. Training that makes one or both parties unhappy is not optimized in terms of functional effectiveness.

LIMA versus LIFE: conclusions

Overall, the LIFE model still requires a caregiver or practitioner to be experienced and skilled when modifying behaviour, but presents a framework to optimize the methods used in terms of both functional effectiveness and the wellbeing of the animal in question. This is certainly the most appropriate practice to follow when a dog is required to partake in active learning. Readers of this book are actively encouraged to read Fernandez's original paper (Fernandez, 2024).

Equipment

An important contributor to a dog's overall wellbeing is the equipment that they are regularly exposed to. Typically, dogs are exposed to these things a couple of times a day when going out for a walk. For some dogs, however, such as those required to wear anti-bark collars inside the home, their caregiver's choice of equipment plays an even bigger part in their wellbeing.

The use of aversive equipment (such as shock and prong collars, slip leads in a garotte position and choke chains) is concerning and demonstrates a widespread form of positive punishment in the general public. The 2024 PDSA PAW Report found that 22% of their sample used at least one aversive method to train their dog (including rattle cans, spray cans and water pistols, as well as the above wearable equipment).

Considering the LIFE method discussed previously, these equipment choices generally violate all three advisories: restricting autonomy, ignoring behavioural function and limiting effectiveness. Sadly, suppressing a behavioural presentation in order to hasten human convenience is all too common. In particular, unqualified practitioners will use multiple aversive methods to compensate for a lack of knowledge and ability with disregard for the welfare of the animal in question.

Ethical considerations

As previously discussed, dogs trained with positive reinforcement have been shown to present with fewer behaviour problems in some research (Blackwell *et al.*, 2008), possibly carrying a lesser emotional burden than those exposed to aversive methods. Therefore, for the purpose of ethical optimization readers are advised not to use slip leads[1], prong collars, vibration, citronella or e-collars.

In the UK, dogs are required by law to wear a collar and tag (Control of Dogs Order, 1992) and so I always ensure that my customers use a traditional flat collar alongside any other equipment. I do not, however, advise attaching the lead to this collar unless the dog walks with little to no tension on the neck. There is little scientific evidence to suggest that dogs walked on flat collars experience greater stress than those walked on harnesses; however, some small studies have recorded a lower ear position in dogs that have a history of being walked on a collar (Grainger *et al.*, 2016). This might indicate that a more negative emotional response has developed over time, but during the study in question the researcher did not record any further evidence to support this.

An equipment choice that often allows a handler to have greater physical control over their dog is a well-fitting headcollar. This should be introduced slowly with positive reinforcement so as to habituate the dog and avoid behaviours such as nose rubbing. A dog that has a suitable, comfortable headcollar and has been positively introduced should not experience an accelerated fear response compared to a flat collar (Ogburn *et al.*, 1998). Not only is this a desirable emotional outcome, but when used safely a headcollar avoids the potential physical complications of a dog pulling against a flat collar around its neck.

As humans, we can select the most well-fitting and ethical pieces of equipment for our dog, but they still might find them aversive. For example, many dogs will still shy away from a comfortable harness. In order to reduce stress for any dog exposed to equipment it is important to introduce them positively right away. If somebody obtains a puppy they should start to gradually introduce equipment even before the puppy is able to go on their first walk. If an individual obtains a rescue dog with a negative response to a certain type of equipment, they should consult a behaviourist and work systematically through the problem. Even the safest and most welfare-friendly equipment choices can be highly aversive if the dog is frightened. Ideally, this work should be prioritized and completed multiple times a day with a high value reinforcer such as food.

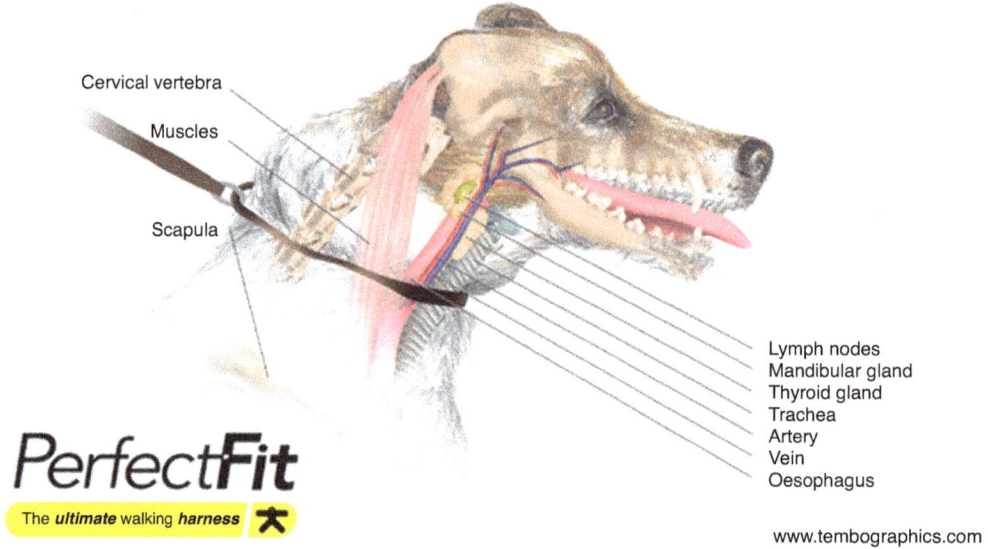

Fig. 6.4. Important anatomical considerations for the dog's neck and the application of equipment such as collars (Dog Games LTD). Used with permission.

Physical considerations

Though it is commonly found that without adequate training a dog is more likely to pull against a harness than a collar (Bailey *et al.*, 2025), there are safety issues associated with only using a collar to walk a dog. Certainly, it is foolish to believe in marketing promising a 'no pull' harness, but a well-fitting harness is optimum when used alongside loose-lead walking training.

When tested on a 'canine neck model' a variety of collar designs (including slip leads) applied at varying pressures found that 'no single collar tested provided a pressure considered low enough to mitigate the risk of injury when pulling on the lead' (Carter *et al.*, 2020). Further still, dogs walked solely on a collar have been proven to be at higher risk of intraocular pressure and so individuals with certain eye conditions are encouraged to use a harness (Pauli *et al.*, 2006). It could be argued that it is safe to use only a collar if a dog has perfect lead walking skills; however, this does not account for unexpected pulling due to triggers or perhaps an accident (for example, a dog falling down a bank or into water). Fig. 6.4 shows the delicate anatomy of a dog's neck.

Electronic collars (e-collars)

Electronic shock collars (otherwise known as e-collars) have been used since the 1930s (Johnson and Wynne, 2024). A controversial piece of equipment, these collars deliver electric shocks to the neck of the dog when a 'command' is failed or if the dog shows an undesirable behaviour. Many countries have banned the use of e-collars due to widespread condemnation of such methods. Despite this, some individuals and organizations claim that these collars are necessary for certain situations, problems or sports, and that they can achieve results that cannot be achieved using positive reinforcement. However, there is no evidence whatsoever to support this claim. Interestingly, a survey of 3897 respondents showed that men are more likely to choose to use an e-collar than women, though the reason why is unclear (Blackwell *et al.*, 2012).

A 2020 study (China *et al.*, 2020) compared training success between e-collar trainers and positive reinforcement trainers in a sample of 62 dogs. This study found that the positive reinforcement group required less verbal and hand signal cues to achieve the behaviour and also showed less delay in response. Further research has tried to validate the use of e-collars but has not been successful due to obvious flaws in experimental design. What said studies *have* successfully proven, however, is that dogs being shocked by e-collars experience a degree of pain or discomfort, with a 2024 study describing how all dogs exposed to shock collars yelped upon being shocked (Johnson and Wynne, 2024). With all of this considered, e-collars are not recommended for use in any situation since there are clear ethical implications and positive reinforcement has been shown to be more effective.

An Optimum State for Learning

As well as utilizing the correct equipment and training methods, certain prerequisites are necessary for optimum learning. Exposure to environmental stimuli as well as a dog's current emotional state are both important considerations.

Arousal and emotional valence

Arousal can be defined as 'physiological and psychological activation into a state of general wakefulness or attention' (Starling *et al.*, 2013). When explaining arousal to caregivers it can be helpful to remind them that we are not necessarily measuring sexual activity, but rather all activity in the body that causes a dog to increase in alertness.

Think of arousal on a scale of temperature, as shown in Fig. 6.5. Different levels of arousal are required for different activities, for example simpler activities can be performed when 'hotter' or more aroused. However, when using positive reinforcement a moderate level of arousal is optimum (Starling *et al.*, 2013). If a dog is threatened, possibly in 'fight or flight' mode, they are much less equipped to behave appropriately and are often unable to concentrate on basic tasks (Mendl, 1999). Caregivers will often describe that dogs in this state are forgetful of known behaviours and unable to accept reinforcers that they usually enjoy. All of these things make training and behaviour modification in a high-arousal situation almost impossible.

Similarly, dogs who are not aroused enough to learn will likely show reduced interest and motivation.

Methods for raising arousal include:

- Exposure to reinforcers such as toys or food.
- Play.
- Changing the environment (for example, moving from a dog's resting space to a different room).

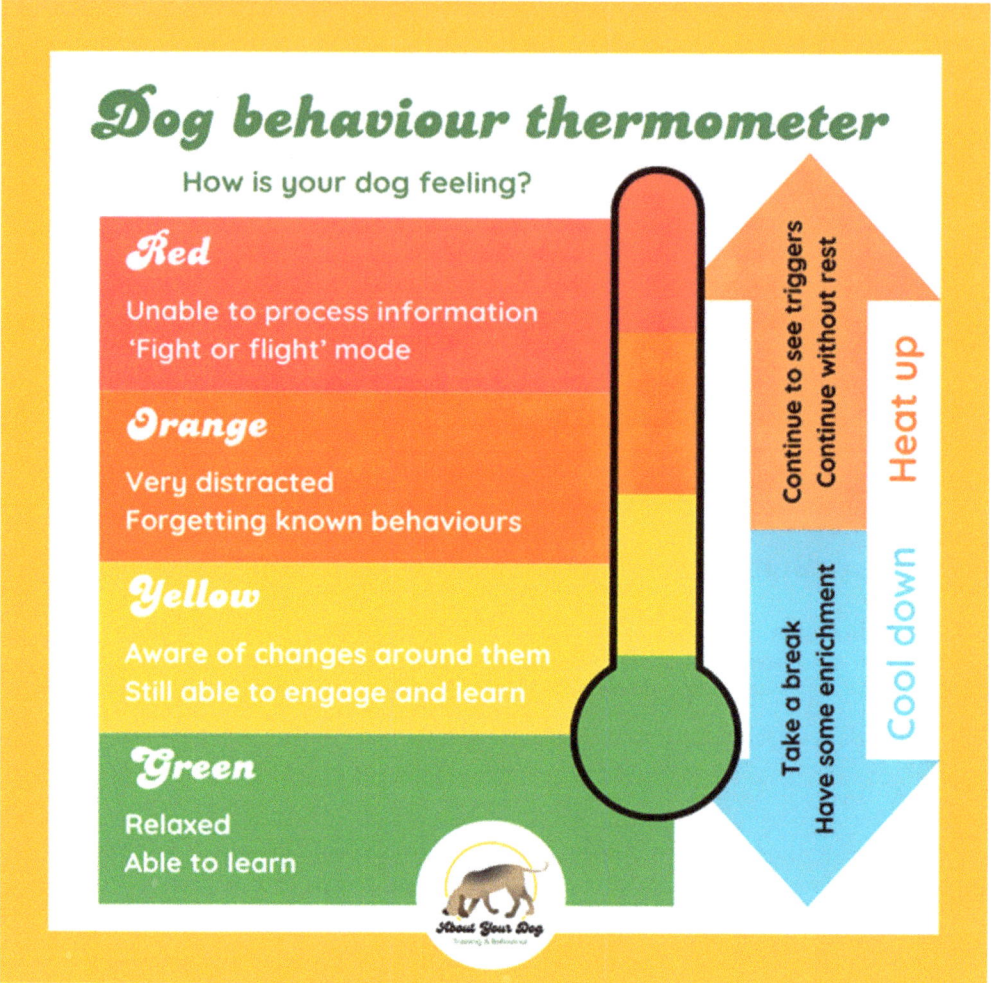

Fig. 6.5. Dog behaviour thermometer. Image is author's own.

Methods for lowering arousal include:

- Encouraging sniffing (either a toilet break or actively setting up sniffing activities).
- Licking and chewing.
- High-quality rest and sleep.

Research across various species also suggests that cognitive bias can be as impactful as arousal when an animal is trying to learn. For example, studies in non-human primates have shown that individuals experiencing depression or anxiety are less likely to be optimistic about learning novel tasks (Bethell *et al.*, 2023). When implementing a training or behaviour modification protocol, it is important to establish whether cognitive bias could be hindering the individual's progress. Would it be beneficial to work on confidence building first?

Trigger stacking

> Trigger-stacking occurs when a dog undergoes multiple stressors without adequate return to baseline. This phenomenon can lead to the appearance of "unpredictable" behavior, such as growling or refusing to engage in an activity; however, many canine stress-related behaviors can be anticipated with thorough understanding of their antecedents.
>
> (Townsend and Gee, 2021)

The accumulation of stress over time can cause a dog to go over their threshold. This refers to the point up to which they can continue to cope. Once over this point, a dog will behave with increased intensity, often acting impulsively or in some cases completely shut down.

When setting a dog up for success, trigger stacking is a major principle to consider (Fig. 6.6). Caregivers and professionals should be aware of stressors that the dog has been exposed to recently and ensure adequate time for sleep, rest or decompression. If the dog is behaving poorly or responding excessively to something, it is always worth considering whether they are too over threshold to learn effectively.

Distractions

As well as internal processes contributing to a dog's ability to learn, external factors have a sizeable impact.

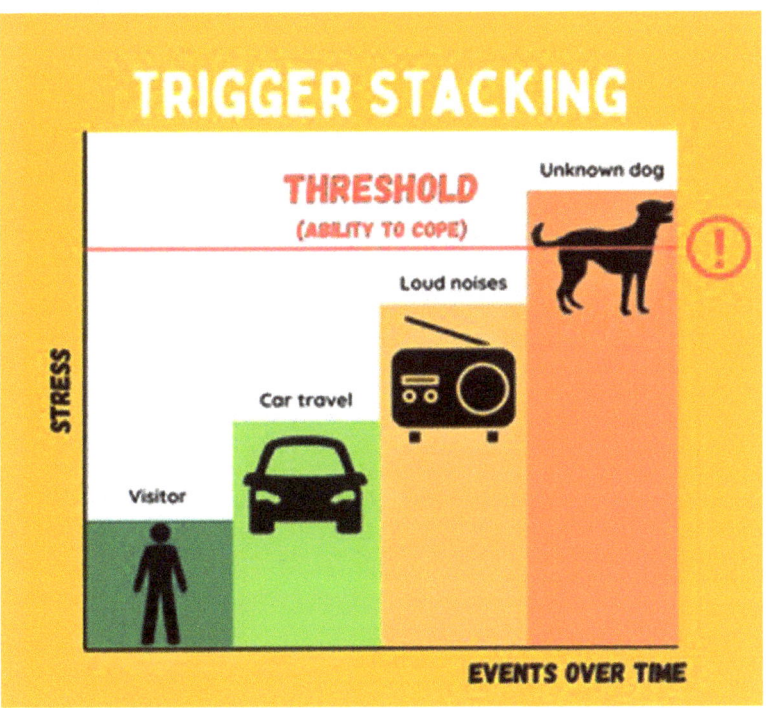

Fig. 6.6. Trigger stacking. Image is author's own.

For example, though training with distraction is important for proofing a behaviour, it is more efficient to teach a new behaviour in silence before gradually introducing novel sound later on (Sheldon *et al.*, 2023). It is worth therefore considering whether visual stimuli might be hindering the early progress of a dog's learning as well as any other external distractions.

Professional dog trainers will be familiar with the '3 Ds of Dog Training':

1. Distance
2. Distraction
3. Duration

Introducing these challenges one at a time and building on them is generally a great framework to follow when teaching new behaviours.

For example, a dog will learn a settle behaviour first and then work on the 3 Ds in turn. The handler may walk away from them (distance), bounce a tennis ball on the floor (distraction) or ask them to stay there for a longer period of time (duration). These practices are key in making behaviours reliable in the 'real world'.

Finally, expecting a dog to learn multiple new things at a time is an unreasonable request. Attending to multiple tasks concurrently will likely cause inefficiency and lead to lapses in your dog's finite attention span (Mendl, 1999). Breaking down a training plan into small, manageable chunks and working on them over an extended period is far more likely to be successful.

Conclusion

Ensuring emotional wellbeing when learning, but particularly through training and behaviour modification, is an area in which so many fall short. In failing to meet these needs, handlers and caregivers risk a breakdown in their mutual bond, inefficient learning and failure to succeed in tasks, creating negative cognitive bias and physical injury among many other things. While some dogs may not outwardly show their unhappiness when exposed to certain stimuli and situations (for example aversive equipment) there is sufficient evidence to suggest underlying discomfort or emotional damage will still take place. Therefore, handlers and caregivers are encouraged to seek out the most ethical training methods, environments and equipment. Additionally, if they are unable to solve problems within the realms of their own knowledge, they should seek further support rather than turning to aversive or coercive methods.

Note

[1] Slip leads can be used as a valid temporary equipment choice for dogs who are equipment-shy or who need to be moved with minimum contact. However, the use of a slip lead as a training device typically relies on the lead tightening around the dog's neck, often where the cervical vertebrae meet the skull. Some 'trainers' will use this to cut off air supply.

References

Control of Dogs Order (1992). Available at: https://www.legislation.gov.uk/uksi/1992/901/contents/made (accessed 14 July 2025).

Bailey, J., Discepolo, D., Baker, J. and Perry, E. (2025) An investigation of force potential against the companion dog neck associated with collar use. *Journal of Veterinary Behavior* 78, 19–24. DOI: 10.1016/j.jveb.2024.10.007.

Bethell, E.J., Holmes, A., MacLarnon, A. and Semple, S. (2023) Cognitive bias in a non-human primate: Husbandry procedures influence cognitive indicators of psychological well-being in captive rhesus macaques. *Animal Welfare* 21(2), 185–195.

Blackwell, E.J., Twells, C., Seawright, A. and Casey, R.A. (2008) The relationship between training methods and the occurrence of behavior problems, as reported by owners, in a population of domestic dogs. *Journal of Veterinary Behavior* 3(5), 207–217.

Blackwell, E.J., Bolster, C., Richards, G., Loftus, B.A. and Casey, R.A. (2012) The use of electronic collars for training domestic dogs: Estimated prevalence, reasons and risk factors for use, and owner perceived success as compared to other training methods. *BMC Veterinary Research* 8, 93. DOI: 10.1186/1746-6148-8-93.

Carter, A., McNally, D. and Roshier, A. (2020) Canine collars: An investigation of collar type and the forces applied to a simulated neck model. *Vet Record* 187(7), e52. DOI: 10.1136/vr.105681.

China, L., Mills, D.S. and Cooper, J.J. (2020) Efficacy of dog training with and without remote electronic collars vs. a focus on positive reinforcement. *Frontiers in Veterinary Science* 7, 508. DOI: 10.3389/fvets.2020.00508.

Collins Dictionary (n.d.) Definition of 'obedience'. Available at: https://www.collinsdictionary.com/dictionary/english/obedience (accessed 14 July 2025).

Fernandez, E.J. (2024) The least inhibitive, functionally effective model: A new framework for ethical animal training practices. *Journal of Veterinary Behavior* 71, 63–68. DOI: 10.1016/j.jveb.2023.12.001.

Grainger, J., Wills, A. and Montrose, T. (2016) The behavioral effects of walking on a collar and harness in domestic dogs (*Canis familiaris*). *Journal of Veterinary Behavior: Clinical Applications and Research* 14, 60–64. DOI: 10.1016/j.jveb.2016.06.002.

Grohmann, K., Dickomeit, M.J., Schmidt, M.J. and Kramer, M. (2013) Severe brain damage after punitive training technique with a choke chain collar in a German shepherd dog. *Journal of Veterinary Behavior* 8(3), 180–184.

Herron, M.E., Shofer, F.S. and Reisner, I.R. (2009) Survey of the use and outcome of confrontational and non-confrontational training methods in client-owned dogs showing undesired behaviors. *Applied Animal Behaviour Science* 117, 47–54.

Johnson, A.C. and Wynne, C.D.L. (2024) Comparison of the efficacy and welfare of different training methods in stopping chasing behavior in dogs. *Animals* 14(18), 2632. DOI: 10.3390/ani14182632.

Lenehan, M. (1986) Four ways to walk a dog: Philosophies of dog training. *The Atlantic*. Available at: https://www.theatlantic.com/magazine/archive/1986/04/four-ways-to-walk-a-dog/667478/ (accessed 14 July 2025).

Mendl, M. (1999) Performing under pressure: Stress and cognitive function. *Applied Animal Behaviour Science* 65, 221–244.

Milgram, S. (1974) *Obedience to Authority: An Experimental View*. Harper & Row, New York.

Ogburn, P., Crouse, S., Martin, F. and Houpt, K. (1998) Comparison of behavioral and physiological responses of dogs wearing two different types of collars. *Applied Animal Behaviour Science* 61, 133–142.

Pauli, A.M., Bentley, E., Diehl, K.A. and Miller, P.E. (2006) Effects of the application of neck pressure by a collar or harness on intraocular pressure in dogs. *Journal of the American Animal Hospital Association* 42(3), 207–211.

PDSA (2023) PDSA Animal Wellbeing (PAW) Report. Available at: https://www.pdsa.org.uk/what-we-do/pdsa-animal-wellbeing-report/paw-report-2023 (accessed 10 July 2025).

PDSA (2024) PDSA Animal Wellbeing (PAW) Report. Available at: https://www.pdsa.org.uk/what-we-do/pdsa-animal-wellbeing-report/paw-report-2024 (accessed 10 July 2025).

Pongrácz, P. (2014) Social learning in dogs. In: *The Social Dog*. Academic Press, Cambridge, Massachusetts, pp. 249–293.

Rehman, I., Mahabadi, N., Sanvictores, T. and Rehman, C.I. (2017) *Classical Conditioning*. Europe PMC. StatPearls Publishing, Treasure Island, Florida.

Sheldon, E.L., Hart, C.J., Mills, D.S., Soulsbury, C.D., Sumner, R. *et al.* (2023) The impact of auditory distraction on learning and task performance in working dogs. *Applied Animal Behaviour Science* 265, 105977. DOI: 10.1016/j.applanim.2023.105977.

Starling, M.J., Branson, N., Cody, D. and McGreevy, P.D. (2013) Conceptualising the impact of arousal and affective state on training outcomes of operant conditioning. *Animals* 3(2), 300–317. DOI: 10.3390/ani3020300.

Todd, Z. (2018) Barriers to the adoption of humane dog training methods. *Journal of Veterinary Behavior* 25, 28–34. DOI: 10.1016/j.jveb.2018.03.004.

Townsend, L. and Gee, N.R. (2021) Recognizing and mitigating canine stress during animal assisted interventions. *Veterinary Sciences* 8(11), 254. DOI: 10.3390/vetsci8110254.

Tunaitytė, K., Ribikauskas, V. and Kučinskienė, J. (2024) Obedience training helps encourage potential owners to adopt shelter dogs. *Journal of Veterinary Behavior* 75, 54–61. DOI: 10.1016/j.jveb.2024.07.001.

Vieira de Castro, A.C., Barrett, J., de Sousa, L. and I Anna, S.O. (2019) Carrots versus sticks: The relationship between training methods and dog-owner attachment. *Applied Animal Behaviour Science* 219, 104831.

Yin, S. (2007) Dominance versus leadership in dog training. *Compendium: Continuing Education for Veterinarians* 29(7), 414–417.

Ziv, G. (2017) The effects of using aversive training methods in dogs—A review. *Journal of Veterinary Behavior* 19, 50–60.

7 Emotional Wellbeing in the Home

Jade Nicholas*

CAB, Winchester UK

Abstract

A safe, secure and predictable home environment is paramount to the emotional wellbeing of any animal. Not only is this part of meeting a dog's basic needs, but the rippling effect of an unstable home environment is hugely impactful for dogs presenting with behaviour problems. A lack of secure bonds, lack of routine or even a lack of decent sleep can be monumental. This chapter explores a dog's relationship with humans (of all ages), the stressors that they are exposed to in their home environment and the impact of unpredictability amongst various other factors. Guidance is given for how to enhance these things and maximize positive experiences from a dog's perspective. It is the intention of this chapter to help caregivers create an optimized home environment and see maximum progress when working through behaviour issues.

Introduction

Given that dogs spend the majority of their life in a single home environment, their emotional wellbeing at home is an incredibly important consideration for their overall welfare. As much as the physical environment is important, the social one carries just as much weight. If dogs are retreating to a space that they do not feel safe in (for whatever reason), it is likely that they are not recovering from stressors in other areas of their life. Therefore, caregivers and practitioners need to optimize a dog's home before anything else.

It is worth acknowledging as well, that a 'safe' home environment is transient. Changes in the environment caused by visitors, building work and other things can quickly make an environment feel unsafe. We invite our children's friends over when we know our dog is fearful of loud noises, or we have family round for a barbecue knowing that our dog struggles to regulate their emotions around food. There may even be factors out of our control such as new neighbours, delivery drivers or building work across the street. Contingencies must be put in place when a dog's experience strays from the norm, not just the 75% of the time when things are static and manageable.

Relationships with Humans

The relationship we foster and the expectations that we have of our dogs are invaluable to their emotional wellbeing and consequently their behaviour. Consistency and predictability are two of the most important contributors to creating and managing a strong bond, but they are frequently found lacking in dog–caregiver relationships. For example, the caregiver that is sometimes kind and affectionate but sometimes aggressive and frightening

---

*Corresponding author: aboutyourdog@outlook.com

will struggle to form a robust bond with their dog and easily work through any ensuing behavioural difficulties.

One study (Brubaker and Udell, 2023) considered 'pet parenting styles' in line with the human literature and established three different styles of 'pet parenting':

1. Authoritative.
2. Authoritarian.
3. Permissive.

An *authoritative* person can be considered to be reliable and trustworthy. *Authoritarian* people are stricter and more willing to restrict freedom, the exact opposite of *permissive* people, who allow freedom with almost reckless abandon. When considering the authoritative personality in dog caregivers (described in the 2023 study as 'high expectations, high responsiveness') research has found that these dogs demonstrate better social skills, secure attachment to people and also greater ability to problem solve than the other two types. Ultimately, neither unfettered liberalism or strict leadership is most efficient in working with dogs, but rather a confident yet fair and predictable caregiver allows a strong social bond and development of important life skills.

As well as the 'parenting style', other factors can impact the dog–caregiver relationship, particularly when considering attachment style. As with humans, attachment style can vary through childhood and in turn have consequences later in life. Dogs and caregivers should seek to nurture a 'secure attachment', characterized by being able to support their dog in a time of need or offering a secure base, as well as giving them the freedom and confidence to pursue other activities unrelated to attachment-seeking (Obeji and Berant, 2010).

A survey of 982 canine caregivers revealed correlations between insecure parenting styles and common behaviour problems. Caregivers categorized as 'disorganized' predicted a likelihood of separation-related problems while those categorized as 'avoidant' predicted more fear-related behaviours. Aggressive behaviour was predicted to be more likely in caregivers categorized as both disorganized and avoidant (de Assis *et al.*, 2025) (Fig. 7.1).

Dogs that have a human to act as a secure base when confronted with stressors show reduced heart rate variability compared to dogs facing things alone (Gácsi *et al.*, 2013). Not only that, experiencing stressors with a secure human beforehand can help dogs to go forward and approach things more confidently on their own. Research has found that dogs shared between other households and also those sharing their home with other dogs might have lesser potential for creating secure bonds (Marinelli *et al.*, 2007). Suggestions of how the secure bond can be achieved include:

- Being consistent in communication style, rules and expectations.
- Regularly enjoy quality one to one time but also ensure that the dog can cope with being alone for short periods.
- Encourage exploratory behaviour and reward for confidence.
- Encourage strong relationships with other animals and people.
- Provide a safe base, particularly when confronted with stressors.

Additionally, dogs have been shown to synchronize with chronic stress in their caregivers over time (Sundman *et al.*, 2019). While it has been noted that this can be a result of physical activity as much as mental, it is still important to consider the effect we are having on our dogs when we are not taking care of ourselves. Therefore, caregivers experiencing complex problem behaviours in their dogs will likely benefit from self-care, and thought for their own emotional regulation.

Dog–human attachment styles	Caregiving style of owner towards their companion animal
Secure (primary strategy)	Flexible: sensitive, reliable, available, supportive; enjoys the relationship
Insecure–avoidant (secondary strategy)	Minimized care or distanced: keeps physical or emotional distance; devaluation of attachment needs; focuses on animal's independence or entrusts care to someone else; might see the animal as requiring authoritarian discipline.
Insecure–ambivalent (secondary strategy)	Heightened care but is uncertain: keeps constantly close and interferes with animal's independence (i.e., exploratory behaviour, especially when away from the owner).
Disorganized (breakdown of primary and secondary strategies)	Defective: intermittently and inconsistently 'abdicate' their protective role; potentially neglectful, abusive or frightening behaviours towards the animal; see themselves as unable to take care of the animal especially during stressful events (owner may feel frustrated).

Fig. 7.1. Dog–human attachment styles in detail (de Assis et al., 2025).

Babies and children

There are many known benefits for children growing up with dogs. Children who share their home with a dog benefit from learning empathy, experiencing a reduction of stress and oftentimes partaking in more physical activity (Giraudet et al., 2022). It is therefore not surprising that many caregivers choose to bring children into their lives, or conversely that families seek the addition of a dog. Unfortunately, it is not unusual for such decisions to bring about difficulty in the home.

Children are a common stressor for dogs. Dogs may experience stress throughout a human's pregnancy, often due to changes in routine and behaviour. Then, a child being born often causes further problems, compromising on sleep (quality and quantity), time spent with their caregivers and for some dogs, the increased presence of visitors.

But, it doesn't stop there! As children continue to grow and develop, they change their movement and communication too quickly for some dogs to cope with. Imagine a nervous dog taking the time to become acquainted with a stationary baby, and then suddenly the scary thing starts crawling! Add into the mix a child's lack of ability to read signals (Davis et al., 2012; Arhant et al., 2016) and a dog trying to communicate that they are uncomfortable is placed under even more pressure. Activities like 'dress up' are common, but are known to be stressful for many dogs (Hall et al., 2019). In many cases, dogs are expected to adapt *overnight* to a completely new way of life, leading to social, psychological and even physical effects (Hall et al., 2019).

Anecdotally, in recent years I have noticed an increase in dogs purchased as 'therapy pets' for children with additional needs. While dogs can be magical support systems for children that are struggling, it is easy for their welfare to become compromised. To provide some common examples: these dogs might be expected to withstand emotional meltdowns, unpredictable behaviour and even aggressive outbursts. For any caregivers or professionals reading this and intending to introduce a 'therapy dog' to a child, please remember the following:

- Dogs cannot be expected to have limitless resilience and capacity for stress.
- Dogs need time for recovery and decompression as much as children do.
- Not all dogs are cut out to fill a therapy role; *in fact, I would suggest that the majority are not.*

Introducing your dog to a baby

There is no perfect way to introduce a dog to a baby (and vice versa), and so this section is prefaced by advising that there is *always risk involved. Further still, if a dog has history of nervousness, aggression or bites this process should be supervised by a Clinical Animal Behaviourist (CAB) or Veterinary Behaviourist (VB).*

Ensuring that the situation has been thoroughly risk assessed by an appropriate professional, the following guidance is useful for preparation followed by initial introductions:

- Practice carrying a doll (or similar) for a few weeks prior to birth.
- Play sound 'baby sounds' at a low volume for a few weeks prior to birth.
- Provide the dog with blankets and clothes that smell of the baby prior to meeting.
- Conduct on-lead introductions first (it may even be favourable to have a baby gate also) whilst all adults are stood up.
- Gradually reduce restrictions in line with advice from a CAB or VB (see above).

It is never safe to leave a child alone with a dog, but it is even more dangerous to leave a baby unsupervised and so this should be avoided *without exception.*

Sleep

An enormous barrier for canine wellbeing is insufficient sleep. As any new parent will tell you, broken sleep is often on the cards with a baby. Therefore, dogs that sleep in proximity to their caregivers (e.g. on the bed) are likely to suffer from the same broken sleep patterns, likely diminishing their welfare.

As a result, prior to bringing a baby home it can be beneficial to encourage a dog to sleep elsewhere, preventing them from being disturbed so frequently. If this is not possible a contingency is to prioritize a dog's access to sleep during the day. One way that caregivers can do this is by sitting calmly for a period of time and encouraging the dog to rest beside them, as continuing to buzz around the house will restrict the dog's ability to switch off.

Distribution of attention

Understandably, new parents typically wish to partake in activities such as bathing and feeding their baby together. For dogs that have not been accustomed to sharing the attention of their humans, it can be incredibly frustrating to be locked out at these times, particularly if it interferes with the part of the day normally assigned to cuddles on the sofa. Where possible, parents can discuss how to divide their time or space in a way that meets the needs of all parties. For example, if one parent is bathing the baby, the other can be sat within proximity enough to feel included but giving the dog a fuss with one hand.

Activity sharing

In order to create a positive association with the new household addition, it can be good to provide enrichment in (safe) proximity to the baby. Using the example above of parents partaking in bath time, the dog can be provided with a lick-mat in the hallway within view of their family.

Secure spaces

As well as providing a 'safe haven' (see Chapter 5), caregivers can manage the shared space of baby and dog to set them up for success. This particular intervention is frequently helpful for avoiding conflict around resources. So, not only does the dog have somewhere they can go to rest or sleep, they also have space extended outwards where they can play, relax or chew undisturbed. A gate closure means that babies and toddlers cannot encroach if the dog is choosing to retreat to their secure space; this can be relaxed as children get older and are able to follow instructions not to cross the boundary (Fig. 7.2).

It is also helpful to consider containing the resources of the child in their own designated playpen, avoiding temptation for curious dogs to start stealing. Alternatively, caregivers might choose to allocate a playroom that is inaccessible to the dog.

Establishing boundaries

Of course, establishing boundaries goes further than physical perimeters. Particularly bold children may be more predisposed to risk taking (Davis *et al.*, 2012) and therefore caregivers will benefit from clear rules and some non-negotiable 'dos' and 'don'ts'. For example, it is likely to be helpful for calm behaviour in children to be prioritized above all else, since dogs are known to synchronize their behaviour with children (Wanser *et al.*, 2021), and dogs can sometimes be a target for aggression when children are angry (Hall *et al.*, 2019). Children allowed to run riot might see that behaviour reflected back at them in their dog. Suggestions for rules and boundaries include:

- Do not touch the dog when he/she is sleeping.
- Do not go into the dog's bed.
- Do not touch the dog's food (or remove food from them).
- If the dog growls or shows his/her teeth, move away.
- Do not roughly handle the dog, including tugging his/her fur or ears.
- Be quiet when around the dog; noisy play should be in another room.

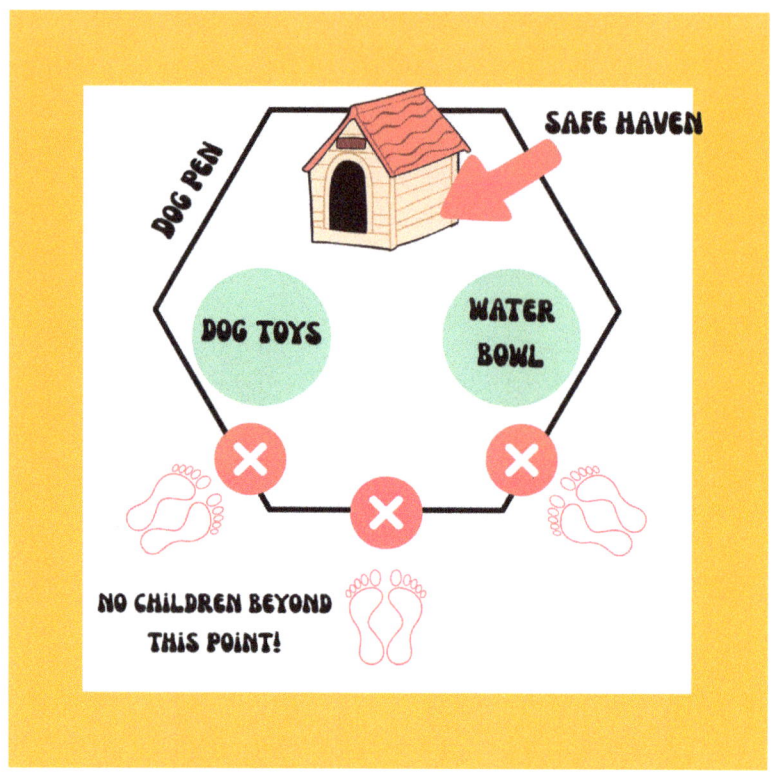

Fig. 7.2. Creating a boundary around your dog's safe haven can provide double security. Image is author's own.

Visitors

It is not uncommon for dogs to struggle with visitors, though the severity of their stress ranges from over-arousal with familiar faces, to those who behave aggressively on sight of anyone not immediately in their social circle. These issues might occur for a variety of reasons and at varying stages in a dog's life. It can be tempting to punish these undesirable behaviours in an attempt to reduce safety risks or social embarrassment. However, many dogs will only see this as confirmation that visitors are scary or stressful. It is much more effective to put time and energy into teaching desirable alternatives and building a dog's confidence slowly.

Environmental management

For dogs that are comfortable in their own company, choosing to remove them from the situation altogether can be an easy way to offload their stress. A safe haven should be implemented; you may choose to do this in a closed room away from the activity.

However, some dogs with more manageable behaviour issues may cope with a safe haven closer to the activity provided that all parties know not to encroach on this private space. A stairgate or room divider can help to enforce these important boundaries.

Practising calmness

As well as having a designated space to relax in, the dog in question will probably need to be taught to behave calmly. The easiest way to achieve this is by providing low-intensity enrichment that encourages chewing, licking or sniffing. Try not to provide anything too active or exciting!

Some dogs will benefit from being reinforced for using a bed and settling down. However, it is important for the individual training the exercise to adopt a relaxed position close to the dog (for example, sat beside them) and reinforce with voice or gentle strokes. Treats or toys are typically highly arousing and may extinguish the desired effect.

Human behaviour

Perhaps the most important intervention that can be made for dogs that struggle with visitors is to manage the behaviour of the visitors themselves. It can be helpful to give instructions ahead of time to confirm what behaviours are expected. Generally, things to avoid include:

- Managing greetings in confined spaces such as doorways or corridors.
- Allowing guests to approach dogs, particularly if they are moving away or behaving defensively.
- Allowing guests to approach a dog in their safe space.
- Allowing guests to reach towards, learn over or crouch closely to dogs.
- Pressuring dogs to interact.
- Wearing hats and sunglasses that obscure facial expression.

A preferable alternative is to wait for the dog to approach at their own speed, perhaps after observing calmly from a distance to begin with. Even at this stage, the dog should lead all interactions.

Touching with consent

A good rule to observe when a visitor is physically engaging with a dog is to regularly test for consent. This exercise is very simple:

1. Stroke (in a safe area away from the face) for 3 s.
2. Remove the hand and wait for the dog's response.
3. If the dog asks for more, perhaps by nuzzling into the person or pawing them, the person can repeat Step 1.
4. If the dog freezes or moves away, the person should immediately stop touching them.

'Treat and retreat'

If the dog is not at the stage that they can tolerate being up close with a visitor, the 'treat and retreat' exercise is easy and effective. Essentially, this exercise teaches that an unknown person approaching yields good things but without compromising the dog's safety or agency.

The exercise looks like this:

1. Begin by settling the dog in a safe space (for example, behind a stairgate).
2. Visitor comes into view and, without looking at or engaging with the dog, throws a treat *behind* the dog.
3. Visitor leaves (moves out of view).

This can be repeated multiple times providing it does not agitate the dog; otherwise one or two repetitions will still have an effect.

Routine and Predictability

One of the most basic things a caregiver can provide for their dog is a predictable routine, in this context a routine in the home. Unpredictable stressors are known to be more stress-inducing than those that are predictable (Dess *et al.*, 1983). Therefore, unexpected bursts of visitors, changes to working patterns or even small adjustments to walk and meal times can be highly stressful for dogs. Avoiding unnecessary routine changes or conversely simply making an effort to keep things predictable will have an impact on the wellbeing of most dogs.

If short periods of unpredictably are expected (for example, building work) caregivers should ensure that they can provide adequate periods of decompression as well as keeping the remaining routine intact. Routine changes contribute to 'trigger stacking' (Chapter 6) and can easily tip an already stressed dog over the edge.

Environmental unpredictability

External triggers (around the periphery of the home) are also unpredictable most of the time, and so environmental adjustments to block their impact will help dogs under pressure, particularly those who are reactive to visual or auditory stimuli. In the National Dog Survey 2024 (Dogs Trust, 2024) in which data were collected for more than 430,000 dogs, 52% of dogs barked at noises outside the front door.

Dogs that are permitted to react repeatedly to visual stimuli, for example, may become hypervigilant to doors and windows, keeping them over-aroused and preventing them from switching off. This is also the case for dogs who are extremely sensitive to sounds. While extreme sensitivity to external stimuli is likely to require professional intervention, *caregivers can take some simple steps including*:

- Closing blinds and curtains, or alternatively using a semi-opaque window cover to block visual access.
- Restricting access to rooms that look out onto 'trigger' areas.
- Consistently having the TV or radio on in the background (or at least at busy times).
- Removing the dog from the house at busy times, perhaps going for a walk to avoid the school run.
- Increasing environmental enrichment to distract and to reduce boredom.

Changes for the better

Although guidance is generally to keep routines consistent, it may be worth reshuffling initially in order to enhance emotional wellbeing in future. In modern society, many caregivers want to take their dogs everywhere with them, but in fact time spent alone in a safe and secure environment can be far more beneficial.

Case vignette: Twiglet

Twiglet is a 17-month-old Jack Russell X Dachshund. Twiglet's caregivers were seeking treatment for her unpredictable aggression towards people along with general reactivity to stimuli around the periphery of the house. Initial assessment revealed that Twiglet would spend most days at work with her male caregiver in a busy workshop environment with intermittent breaks for walks, during which time she would see lots of people and dogs in a fairly restricted space. Once home she would spend multiple hours showing hypervigilance to the main window looking out onto a main road, reacting as people and dogs walked past.

If left alone Twiglet was relaxed and would switch off. Therefore, it was decided that Twiglet's caregivers would trial leaving her alone at home during the working day (with regular pop-in breaks), also meaning that she could be walked in an open space closer to home. Blinds over the main window were also utilized to restrict Twiglet's visual access.

Fig. 7.3. Twiglet now enjoys her time alone as well as time spent with visitors. Image is author's own.

Given Twiglet's combination of breeds, she is by nature a vocal dog and will still vocalize at the doorbell and on initial greeting. However, after 6 months Twiglet's hyper-arousal and stress-related behaviours are enormously reduced. Consequently, Twiglet has reduced significantly in reactivity to dogs and people as well as slowly being able to welcome guests into her home (Fig. 7.3). Twiglet's caregivers commented that closing the blinds initially had an overnight impact on her behaviour.

Leaving a Dog Alone

The 2024 PDSA Paw Report revealed that 11% of dogs of 2371 owners surveyed (approximately 260 dogs) showed signs of distress (including vocalizing and destructive behaviour) when left alone (PDSA, 2024). 9% of dogs recorded in the Dogs Trust National Dog Survey 2024 (approximately 38,700 dogs) were unable to relax when left alone. These numbers do not account for dogs who are not monitored when left alone, and may be exhibiting signs of distress without their caregiver realizing. Some earlier studies have estimated that over 20% of the general dog population exhibit problem behaviours related to being left (de Assis *et al.*, 2020).

Many treatment programs for problem behaviour benefit from the dog spending time alone, whether it is to allow the dog to decompress or to help them avoid exposure to their triggers. Therefore, it is not unreasonable to suggest that having the option to leave a dog on their own might positively influence their wellbeing.

Recent research has determined multiple possible reasons that a separation related problem (SRP) might occur, and so caregivers are unlikely to be able to work through these presenting complaints alone. There is, however, a risk that some practitioners set in their ways will approach these problems with a 'one size fits all' approach, discussed at length in a 2020 paper by de Assis *et al*. Though all dogs in this scenario experience distress when left alone, there are multiple compounding factors that dictate how they will behave.

The 2020 paper proposes four main types of SRP for practitioners to diagnose and then treat accordingly:

1. **Exit frustration:** Behaviour (typically destructive) directed mostly to exits
2. **Redirected reactive:** Agonistic to stimuli outside of the environment, redirecting frustration inside due to a lack of access.
3. **Reactive inhibited:** Reactive to stimuli outside but without trying to get access (possibly overall more fearful or avoidant).
4. **Boredom related:** A lack of stimulation builds frustration and sometimes redirection.

Ultimately, a secure bond with a caregiver, solid routine and enriched environment will reduce the risk of SRP in many dogs, and for some these interventions will be enough to eliminate problem behaviour. However, for those that demonstrate more complex behaviours categorization into the type of SRP and subsequent treatment with a CAB or VB is highly valuable.

Relationships with Other Animals

Dogs that live with other animals, whether conspecifics (other dogs) or other species, can experience difficulty. Some will experience occasional disagreements, while others will be put under daily pressure by conflict and even relationship breakdowns. Not only is this primary relationship a stressor for that dog, but inevitably humans will become frustrated and emotional when trying to manage it.

Temporary disagreements may be safely negotiated by the animals themselves, but caregivers will likely need to intervene and rectify ongoing conflict in order to enhance wellbeing in their pets.

Dogs living with other dogs

Although many dogs benefit from being housed with others (Hecker *et al.*, 2024), it is not always straightforward to manage a multi-dog household. Conflict can arise for a variety of reasons including disagreements over resources, differences in energy levels (for example one individual wanting to play) or even irritability in either party caused by illness or injury.

A review of 38 cases (Wrubel *et al.*, 2011) established some key trends in inter-dog aggression in the home:

- Fights between dogs in the same household tend to result in worse injuries than those in separate households.
- Typically there is a primary instigator, and this individual is usually younger or newer to the household.
- Changes in the household are a common trigger for behaviour change, including a dog passing away, a dog becoming sick or a dog maturing sexually.

Previous studies determined that females were more likely to instigate fights within the household than males, and that the number of fights was greater between same-sex pairs (Sherman *et al.*, 1996).

Common antecedents for incidents occurring include disputes over resources, competition for human attention, and conflict in confined spaces. Alternatively, excitement can also quickly lead to aggressive incidents (Sherman *et al.*, 1996; Wrubel *et al.*, 2011).

The prognosis for these problems needs to be guarded, as some relationships cannot recover and dogs will need to be kept separately indefinitely. Often, the prognosis is worse if the instigator of the fights is the younger dog, if humans have been bitten, or if incidents are thought to be unpredictable (Wrubel *et al.*, 2011).

Questions regarding welfare should be raised immediately when dogs are unable to co-exist at all together. Compromises to safety, autonomy and freedom of expression are very costly to the emotional wellbeing of a dog, and so the question of possibly rehoming one party is necessary to raise early on. To give an example, if one dog is frightened of being attacked every time they walk into a room they cannot be expected to live under that pressure for an extended period.

Despite some difficulties likely occurring during treatment, it is always worth attempting some treatment in order to restore harmony. Anecdotally, dogs that continue to show affiliative behaviour (grooming, resting together and playing) towards one another are much more likely to show reduced aggression. Otherwise, if dogs are ambivalent towards one another or worse, are aggressive on sight, treatment is far more challenging.

Separate spaces

It is valuable to offer each individual a separate safe space. This might be as simple as providing multiple crates, or a split in the home environment might be required.

These separate spaces mean that dogs can be settled apart from one another at times when triggers are likely to be around. For example, a classic (and simple) remedy for food aggression between dogs is to feed them in separate rooms. This also provides one dog with the opportunity to retreat from the other when wanting to avoid conflict.

The 2011 study (Wrubel *et al.*, 2011) found that half of the dogs presented with a degree of separation-related issues. This could possibly lead to an over-dependence on their caregiver and therefore a need to try and gain priority access to them. So, the use of an allocated safe space might also help to build some independence and a more secure attachment to caregivers in the home.

In the instance a fight has occurred it is generally worth separating the dogs temporarily (if safe to do so) until their emotional arousal has lowered. Depending on the layout of the house and the ability to access resources, this can be temporary use of small spaces, or this can be permanent divisions of the home environment. Even if the dogs are required to be separated for a longer period of time, these separate spaces can help desensitize one another to the presence of their conspecific.

From the safety of these individual spaces, dogs can be given enrichment to promote calm behaviours. Observing one another in a passive non-confrontational way is incredibly beneficial over time as they begin to see their counterpart as non-threatening.

Individual resources

Though many dogs will happily share some resources by choice, basic resources should always be provided per number of dogs. In some circumstances, though multiple resources are provided, dogs will try to steal from one another, which can easily lead to conflict. In these instances, the resources in question should be provided in the separate spaces.

Named feeding can also reduce anticipatory conflict between dogs when waiting for a reward. Rather than throwing food rewards and risking dogs fighting amongst themselves, naming the dog as you reward them (and repeating for however many dogs you are working with) can establish boundaries regarding what belongs to whom. It is known that dogs are sensitive to inequity in rewards (Wrubel *et al.*, 2011) so those using this exercise should make sure that they distribute rewards equally.

Appropriate play

Given that excitement is a known trigger for aggressive outbursts between conspecifics, restrictions should be in place for high arousal activities or incidents.

One example of a positive restriction is a time limit for play, avoiding dogs from becoming too over-aroused and crossing into aggression. With a break for cooling down, dogs might be able to then revisit play in short bursts again.

It is always valuable to moderate play to discourage any unwanted behaviour that might lead to retaliation. For example, some dogs may nip, grab limbs or otherwise play slightly too roughly. If it looks like the other dog needs a break or to escape from the aggressor, it is worth safely separating them (perhaps using a positive interrupter) to prevent this play from escalating.

'Resetting the room'

Since conflict between dogs can commonly be initiated by one coming into an area that has been 'claimed' by the other, a 'reset' method is a great idea. Anecdotally, it is not uncommon for one dog to be resting with a caregiver and then show aggression to another when they try to approach. If this is a frequent occurrence, it is advisable for the caregiver to stand up and leave the room as the second dog approaches before returning again after a few moments, therefore resetting the established dynamic of the space.

The relationship between pain and behaviour is discussed at length in Chapter 9, but it is not uncommon for dogs that show aggression from a resting position to be dealing with underlying chronic pain. So, caregivers and practitioners are advised to rule this out as a priority alongside the aforementioned interventions.

Dogs living with cats

A common companion for the domestic dog is the domestic cat (*Felis silvestris catus*). Though it is long established that dogs and cats are 'mortal enemies', many households choose to keep both pets. In fact, dogs and cats can live very happily together when set up for success. It is important to consider the emotional wellbeing of both parties when integrating them into the same home; it is not adequate simply to exclude them to separate parts of the home and not consider the implications of this.

Environmental modification

One of the most valuable things people can do to successfully integrate cats and dogs into the same home is to modify the environment to suit them both.

As adept climbers, cats will often choose to observe their surroundings from an elevated area (Stella and Croney, 2016). Creating space higher up in the house through the use of shelves, bookcases, cat trees and other pieces of furniture can allow a cat to move through the house freely without worrying about coming into contact with the dog. In turn, both parties can learn more about one another through simple observation instead of being directly confronted. If a cat is required to move around on the floor at the same height as the dog, there is greater possibility of a chase ensuing and possibly injury.

In all homes, but particularly in those housed with dogs, cats must be provided with a secluded and safe sleeping place that is freely accessible to them (Stella and Croney, 2016). It may also be suitable for cats to be provided with outdoor access without needing to go past a dog to get there, for example a window that overlooks a ledge. Finally, basic resources such as food and water must also be fully accessible without needing to go past the dog.

The above considerations also apply to dogs, as it is not impossible to see dogs who are afraid of cats! Therefore, ensuring that the dog is also able to access their resources without conflict is equally important.

Shared calmness

Encouraging the practice of calm behaviours is very effective, particularly when used in line with the environmental modifications discussed previously.

A very simple way to encourage calmness is to provide both parties with calming enrichment such as a snuffle mat (yes, cats can use them too!). Encourage the cat to use their given vertical space, in particular to enjoy their calming activity up there. Both parties then have the opportunity to practise being calm in the presence of one another from the security of their safe space. Redirect any attempt to initiate play or become excitable back to the activity provided. If either party finds this too difficult, they are probably too close to one another and the exercise needs to be progressed at a slower speed.

At some point it may be safe for certain households to introduce both animals a little more closely. If this is the case, it is still advisable to have precautions in place (e.g. a physical barrier or keeping the dog on the lead) and to separate the two if they stray from the calm behaviour previously established.

Dogs living with other animals

There are many other animals that dogs are expected to live with, some more controversial than others. Social media has shown us that it is not impossible for even the most unlikely partnerships to become friends; however, this cannot be expected. Considerations for dogs living with other animals should include:

- Does the dog have experience of living with this animal previously? Was it successful?
- Does the dog have a strong desire to chase, catch or kill?
- Can the animals be safely kept separate?
- Is the welfare of the second species compromised living in close proximity to a dog?
- Overall, is my dog calm enough to be sensible around this animal?

With these things in mind, accommodations should be made for both animals to have all their needs met and introductions (if appropriate) should be conducted slowly over an extended period of time.

Indoor Environment/Safe Space

Although previously discussed with regards to traumatised dogs, a 'safe haven' is valuable to *all* dogs. So, it should go without saying that when confronted with any problems raised in this book, a safe haven should be provided.

Enrichment

Throughout this book, environmental enrichment is mentioned a lot. Essentially, environmental enrichment is anything that enhances the lived experience of an animal in captivity (Newberry, 1995), and whether we like it or not, our dogs fall into that selection of animals. Enrichment is provided to a multitude of animals over a variety of settings including pet shops, zoos and laboratories and it is thought to enhance both physical and mental wellbeing. For example, studies have shown improvements in cognitive ability in animals provided with enrichment (Petrosini *et al.*, 2009), as well as a marked decrease in stress-related behaviours (Zilocchi *et al.*, 2018). Though toys, commonly provided for this reason, can be enriching, there is a lot more to enhancing the lives of our dogs through enrichment. Offering an animal something functionally relevant is far superior to something provided simply for fun!

Though a generally helpful guideline is to provide something that encourages natural behaviour patterns, each individual is different and so will require a personalized approach to their enrichment (Coleman and Novak, 2017). A caregiver might consider varying levels of difficulty for certain dogs, or in fact different activities altogether! The important thing is to discover what motivates a dog and to use that in their enrichment, as well as understanding that individual's tolerance for frustration so as not to make the task too difficult for them. Studies on non-human primates suggested the following ways to provide enrichment:

- Alterations (levels, textures, flooring, etc.) to the animal's direct environment.
- Visual stimulation.

- Tactile stimulation.
- Auditory stimulation.
- Taste.
- Cognitive stimulation.
- Exercise.
- Social housing (not appropriate for every dog).
- Human interaction.

(Coleman and Novak, 2017)

With all of this considered, enrichment can be provided to reduce boredom, build confidence and also promote calmness. It is good practice to have multiple examples pre-prepared for easy access and ideally to provide them at least once a day. Simple ways to do so include:

- Provide long lasting chews.
- Freeze lick mats or slow feeders.
- Snuffle mats.
- Destruction boxes (typically a cardboard box stuffed with newspaper).

Always ensure that enrichment has been assessed for risk before providing a dog with it. For example, the above activities are safe if given to a dog who will not consume the materials; however, they will not be safe for all dogs. It is always advised to observe the dog in question and intervene if the activity is used inappropriately.

Conclusion

Dogs spend the majority of their time in their home environment, and as such it is important to provide them with somewhere safe and predictable that meets all of their needs. As mentioned earlier in the chapter, we are required to do so (to the best of our ability) 100% of the time, not just when the environment is functioning as normal. Visitors, newborn babies and other pets are common stressors for dogs and will often cause disruption. For this reason caregivers must take the time to implement strategies that support their dog, rather than leaving them to their own devices and expecting them to cope.

References

Arhant, C., Landenberger, R., Beetz, A. and Troxler, J. (2016) Attitudes of caregivers to supervision of child–family dog interactions in children up to 6 years—An exploratory study. *Journal of Veterinary Behavior: Clinical Applications and Research* 14, 10–16. DOI: 10.1016/j.jveb.2016.06.007.

Brubaker, L. and Udell, M.A.R. (2023) Does pet parenting style predict the social and problem-solving behavior of pet dogs (*Canis lupus familiaris*)? *Animal Cognition* 26, 345–356.

Coleman, K. and Novak, M.A. (2017) Environmental enrichment in the 21st century. *ILAR Journal* 58(2), 295–307. DOI: 10.1093/ilar/ilx008.

Davis, A.L., Schwebel, D.C., Morrongiello, B.A., Stewart, J. and Bell, M. (2012) Dog bite risk: An assessment of child temperament and child-dog interactions. *International Journal of Environmental Research and Public Health* 9(8), 3002–3013. DOI: 10.3390/ijerph9083002.

de Assis, L.S., Matos, R., Pike, T.W., Burman, O.H.P. and Mills, D.S. (2020) Developing diagnostic frameworks in veterinary behavioral medicine: Disambiguating separation related problems in dogs. *Frontiers in Veterinary Science* 6. DOI: 10.3389/fvets.2019.00499.

de Assis, L.S., Georgetti, B., Burman, O., Pike, T.W. and Mills, D.S. (2025) Development of a dog owner caregiving style scale (Lincoln Owner Caregiving Questionnaire, LOCQ) and its relationship with behaviour problems in dogs. *Applied Animal Behaviour Science* 287, 106628. DOI: 10.1016/j.applanim.2025.106628.

Dess, N.K., Linwick, D., Patterson, J. and Overmier, J.B. (1983) Immediate and proactive effects of controllability and predictability on plasma cortisol responses to shocks in dogs. *Behavioral Neuroscience* 97(6), 1005–1016.

Dogs Trust (2024) Welcome to the results of the National Dog Survey 2024. Available at: https://www.dogstrust.org.uk/downloads/Dogs_Trust_NDS_Report_2024__.pdf (accessed 10 July 2025).

Gácsi, M., Maros, K., Sernkvist, S., Faragó, T. and Miklósi, Á. (2013) Human analogue safe haven effect of the owner: Behavioural and heart rate response to stressful social stimuli in dogs. *PLoS ONE* 8(3), e58475. DOI: 10.1371/journal.pone.0058475.

Giraudet, C.S.E., Liu, K., McElligott, A.G. and Cobb, M. (2022) Are children and dogs best friends? A scoping review to explore the positive and negative effects of child-dog interactions. *PeerJ* 10, e14532. DOI: 10.7717/peerj.14532.

Hall, S.S., Finka, L. and Mills, D.S. (2019) A systematic scoping review: What is the risk from child-dog interactions to dog's quality of life? *Journal of Veterinary Behavior* 33, 16–26. DOI: 10.1016/j.jveb.2019.05.001.

Hecker, G., Martineau, K., Scheskie, M., Hammerslough, R. and Feuerbacher, E.N. (2024) Effects of single- or pair-housing on the welfare of shelter dogs: Behavioral and physiological indicators. *PLOS One* 19(6), e0301137.

Marinelli, L., Adamelli, S., Normando, S. and Bono, G. (2007) Quality of life of the pet dog: Influence of owner and dog's characteristics. *Applied Animal Behaviour Science* 108(1–2), 143–156. DOI: 10.1016/j.applanim.2006.11.018.

Newberry, R.C. (1995) Environmental enrichment: Increasing the biological relevance of captive environments. *Applied Animal Behaviour Science* 44, 229–243.

Obeji, J.H. and Berant, E. (2010) *Attachment Theory and Research in Clinical Work with Adults*. Guilford Press, New York.

PDSA (2024) *PDSA Animal Wellbeing (PAW) Report*. Available at: https://www.pdsa.org.uk/what-we-do/pdsa-animal-wellbeing-report/paw-report-2024 (accessed 10 July 2025).

Petrosini, L., Mandolesi, L., Giuseppa Leggio, M., Cutuli, D., Gelfo, F. *et al.* (2009) On whether the environmental enrichment may provide cognitive and brain reserves. *Brain Research Reviews* 61(2), 221–239.

Sherman, C.K., Reisner, I.R., Taliaferro, L.A. and Houpt, K.A. (1996) Characteristics, treatment, and outcome of 99 cases of aggression between dogs. *Applied Animal Behaviour Science* 47, 91–108.

Stella, J.L. and Croney, C.C. (2016) Environmental aspects of domestic cat care and management: Implications for cat welfare. *Scientific World Journal* 6296315. DOI: 10.1155/2016/6296315.

Sundman, A.S., Poucke, E., Svensson Holm, A.C., Faresjö, Å., Theodorsson, E. *et al.* (2019) Long-term stress levels are synchronized in dogs and their owners. *Scientific Reports* 9(1), 7391. DOI: 10.1038/s41598-019-43851-x.

Wanser, S.H., MacDonald, M. and Udell, M.A.R. (2021) Dog–human behavioral synchronization: Family dogs synchronize their behavior with child family members. *Animal Cognition* 24, 747–752.

Wrubel, K.M., Moon-Fanelli, A.A., Maranda, L.S. and Dodman, N.H. (2011) Interdog household aggression: 38 cases (2006–2007). *AVMA Journal* 238(6), 731–740.

Zilocchi, M., Casagli, D. and Romano, F. (2018) Problem solving games as a tool to increase the well-being in boarding kennel dogs. *Dog Behavior* 4(1), 9–19. DOI: 10.4454/db.v4i1.v3i3.78.

8 Emotional Wellbeing Outdoors

JADE NICHOLAS*

CAB, Winchester UK

Abstract

This chapter addresses common behaviour problems experienced on walks and when travelling by car. In modern society, many dogs are deprived of their ability to perform natural behaviours such as sniffing, digging and playing freely with others, which ultimately leads to decreased emotional wellbeing. Creating a sense of safety and increasing opportunity to explore enables dogs to become more confident and helps them to relax, in turn enabling caregivers to introduce training and behaviour modification that works. These changes are paramount for dogs that present with reactivity to dogs or people, have a tendency to refuse walks or are fearful of car travel.

Introduction

For most people, dog walks are a fundamental draw of bringing a dog into the home. In the National Dog Survey (Dogs Trust, 2024) 89% of people said that having a dog made them more physically active. However, the time a dog spends outdoors is hugely variable.

One study observing 104 dog–caregiver dyads found that dogs living in flats were more likely to be walked for upwards of 30 min, while 30% of dogs living in family homes with a garden were never walked (Marinelli *et al.*, 2007). Given the multiple benefits (both mentally and physically) experienced by a dog on their walk, this statistic is hugely worrying. It begs the question, do some of us avoid walking our dogs because we are worried about how they will behave? Or further still, do we worry that we are not equipped to make decisions to keep them and the other dogs around us safe?

Since the time spent outdoors is so much less predictable than the controlled environment inside our homes, it is not surprising that many caregivers find walks terrifying. Consider this alongside the lack of consistent advice available on modern training methods, or expectations regarding walking etiquette and the pressure on all parties becomes overwhelming. Perhaps this is why many people resort to stringent restrictions on their dog's autonomy using unforgiving equipment such as slip leads or figure-of-eight headcollars (for more on this see Chapter 6).

This chapter is truly at the heart of this book as it aims to remind caregivers and practitioners how a dog would normally behave in an outdoor environment where human convenience no longer overshadows emotional wellbeing.

Exhibiting Natural Behaviours

One thing immediately springs to mind when considering emotional wellbeing outdoors, and that is a dog's opportunity to exhibit natural behaviour. Not only is this fundamental

*Corresponding author: aboutyourdog@outlook.com

need listed in the Animal Welfare Act (2006), skilled behaviourists will know that embracing the expression of natural behaviour is a key part of most behaviour modification treatment plans. Particularly prevalent with working breeds placed into pet homes, the prevention of natural behaviours is incredibly frustrating for our dogs (Mason and Burn, 2011) and will often be directed onto unsuitable outlets (you may be familiar with collies 'herding' traffic).

Sniffing

The first natural behaviour that is imperative for the wellbeing of all dogs is sniffing. It is well known that the olfactory capability of the domestic dog is extraordinary, resulting in dogs being used for disease detection, forensics, the search for narcotics and of course the tracking of humans. It therefore seems a shame that so many dogs are limited to the boundaries of their back garden, or to a short familiar walking route that they experience every day. Further still, it is an even greater shame that some humans find it such an inconvenience when their dog wants to stop and sniff.

The smells decorating an environment tell a detailed story about the 'comings and goings' of previous days. Through stopping to engage with smells, dogs are able to identify the species, age and gender of animals before them, and it is thought that they can also determine the emotional state of the animal (Kokocińska-Kusiak et al., 2021). For this reason, many people like to equate the dog's experience of sniffing to that of reading the newspaper or checking emails. The dog is catching up on information from that day.

While stopping to let a dog sniff may make the walk longer in duration, or shorter in distance, it is a valuable practice. It does not decrease engagement or demonstrate a lack of discipline. In a human-centric world where dogs have been brought along as passengers it is only fair to let them process it at a speed that works for them.

Canine predatory sequence

To fully understand what 'natural behaviour' might mean to an individual dog, one has to understand the 'canine predatory sequence' (CPS). This description of canine behaviour provides an understanding for everything from play style to the occurrence of problem behaviours. Failing to understand at least the basics of this sequence will leave caregivers with a lot of questions unanswered. In the simplest terms, the CPS is as follows:

1. Orient[1]
2. Give Eye
3. Stalk
4. Chase
5. Grab/Bite
6. Dissect
7. Consume

Different individuals and breed groups will favour certain sections of the sequence. For example, a sighthound is particularly adapted to the 'chase' section but may show less interest in 'dissect' and 'consume'. Pointing dogs may show accelerated ability in 'stalking' but might leave the chase to their faster relatives.

As well as breed expression, different activities provided by humans are designed to meet certain sections of the CPS. For example, offering a destruction box or sacrificing a stuffed toy can meet the need to 'dissect'. Many of the activities listed throughout the next section are designed to meet a certain part of the sequence.

Breed-specific behaviours

Though most dogs can be trained to partake in anything you want them to (with varying levels of success and enthusiasm), there are naturally areas that their breed will have them excel in. Selective breeding has produced dogs with unique abilities to complete certain tasks, meaning that working dogs are very popular. When placed in 'pet dog' homes though, many of these dogs become very frustrated when not able to perform behaviours that are so innately important to them, and this may be where many caregivers report boredom-related behaviour problems. Therefore, one piece in the behaviour puzzle is offering an outlet for breed-specific behaviours.

Scent work and retrieving

Dogs and humans may have been participating in scent work while hunting as far back as 12,000 years ago (Baldursdóttir, 2024). Through a dog's incredible ability to use their nose, they have shown proficiency in searching for narcotics, cadavers and even disease. It is no wonder, therefore, that a dog without cause to use their nose can become easily bored.

A simple game of 'find it' can be set up by throwing some treats into long grass (or even hidden indoors) and prompting the dog in question to search. This can be increased in difficulty over time as the dog becomes more proficient. For those that are particularly interested in progressing their skills, there are scent-work classes such as those provided by *Scent Work Academy* (Scent Work Academy, n.d).

Often, dogs that enjoy scent work are also excellent retrievers. These dogs were bred to traverse rugged landscapes and retrieve game without damaging it (Paddock, 2023). So, it should be no surprise how many of these dogs become fixated with the game of fetch (more on this later). Something often forgotten is the discipline with which working dogs must work in order to retrieve to standard, and this is not replicated in a haphazard game of fetch with a ball thrower. Instead, teaching a reliable 'wait' before tasking a dog with a retrieval of a ball (or hunting dummy) from a bush, tree trunk or from behind a bench requires them to use their skills. Of course, there are plenty of variations to this game for those that are excelling past entry level, and there are plenty of specialist books for caregivers to follow.

Herding dogs

Border Collies are a great example of a working breed that require adequate mental stimulation. When placed in 'pet dog' homes their innate need for herding often does not diminish, and so practitioners frequently see dogs looking for replacement 'sheep' to manage. This may be something relatively harmless like the gentle herding of family members within the home, but can also be incredibly dangerous in circumstances such as frustrated reactions to the movement of bikes and traffic.

Trainer and working dog expert Kay Laurence designed a game called 'Sheepballs®' (The Border Collie Trainer, 2023). This simple game can be taught by caregivers or practitioners and is a fantastic outlet for herding dogs that are not being used for work. Dogs are provided with a selection of balls to represent sheep, and as the caregiver gently moves the ball the dog is tasked with 'herding' the balls back into one group.

Tracking with hounds

People might be scared to lean into the abilities of a tracking dog, assuming that it will lead to terrible behaviour on lead or a complete absence of recall. Anecdotally, due to sharing my own home with a bloodhound, I have found the total opposite. Leaning into a dog's natural abilities in a controlled way may stop them from taking themselves off for other opportunities.

It is thought that dogs such as the Bloodhound were bred for hunting' as early as 2000–1000 BC (Lowe, 1981). Nowadays they are used for a sport known as 'Hunting the Clean Boot', tracking the scent of live humans and indicating when they are found (Fig. 8.1).

Fig. 8.1. Practising 'Hunting the Clean Boot' with Cheddar has undoubtedly deepened our bond, and as a result his engagement on walks is fantastic. Image is author's own.

For those who do not live with a Bloodhound, or otherwise do not wish to enter working trials, 'Man-Trailing' is a great alternative. 'Man-trailing' as an example is an activity that many dogs will enjoy (not just hounds). This modern adaptation of the methods used in search and rescue is accessible in towns as well as rural areas and simply requires the dog to follow a scent trail and locate a 'lost' person.

Keeping terriers busy

After a long period of success in the USA, the game of 'Barn Hunt' is now available in the UK (Barn Hunt International, 2025). In a secure ring of straw bales, dogs locate tubes with the smell of a deceased rat.

Terriers in particular are incredibly driven to search for and kill rodents. In the absence of adequate mental stimulation, this breed group can become fixated on areas where they have previously seen wildlife. For example, garden decking and sheds might become a target, or ponds with fish and mosquitos can become an obsession. As well as plenty of environmental enrichment at home, Barn Hunt (or similar) is perfect to scratch that itch!

Fun for Everyone

Dogs were not designed to be immaculately clean, decorative fixtures in our homes. Though every dog is different, swimming, digging and mud-wallowing are favourite activities among many breeds.

Taking swimming as an example, many large and giant breeds will take every opportunity to spend time in water. Given that large dogs are often predisposed to musculoskeletal conditions, swimming can be a great activity for them to exercise without applying additional pressure to their bodies. In fact, the health benefits of swimming are widely documented in humans and are becoming increasingly documented in dogs. Findings include improved functionality of osteoarthritic joints (Nganvongpanit *et al.*, 2014) as well as weight loss and cholesterol reduction (Nganvongpanit *et al.*, 2016). Caregivers should be vigilant about where they allow their dogs to swim given the possible risks (blue–green algae, stagnant or contaminated water and powerful currents to name a few), but it would be regrettable to prevent swimming altogether with so much to be gained.

Digging and mud-wallowing are also perfectly natural (and enjoyable) behaviours for dogs, but for obvious reasons humans will frequently prevent these behaviours. It is true, digging in particular can occur as a result of discomfort or behaviour problems and so caregivers should seek support if there is sudden onset or things seem obsessive. However, dogs who enjoy engaging in either activity from time to time should not be denied an outlet for doing so.

The Dangers of 'Fetch'

'Fetch' is among the most commonly played games between dogs and their caregivers (Rooney and Bradshaw, 2003). One study reports that 77.8% of participants engage in fetch with their dogs at least occasionally (Delgado *et al.*, 2024). While the physical benefits of exercise are well established, fetch is a game that can easily be overused. Often, this overuse stems from good intentions; however, fetch is frequently used as a convenient alternative by those who opt for repetitive ball throwing, often using devices like ball launchers. Caregivers may refer to their dogs as 'addicts' or 'ball-junkies' without fully acknowledging the implications of these terms.

The nature of the game has also changed over time. What was once a calm back-and-forth activity has become a high-intensity pursuit, with dogs sprinting at top speed after balls flung across long distances, often running until reaching the point of physical exhaustion (Sharkey, 2025).

Such repetitive, high-intensity exercise places considerable stress on the musculoskeletal system. Given the strong link between pain and behavioural issues in dogs (Mills *et al.*, 2020) (see Chapter 9) and the fact that dogs are often willing to endure discomfort for activities they enjoy (Mills *et al.*, 2024), excessive fetch may inadvertently contribute to the worsening of behaviour problems, including aggression.

Fetch is also a game that elicits high levels of emotional arousal in most dogs. This can be observed physiologically, through signs like dilated pupils and heavy drooling, and behaviourally, via actions such as barking, spinning or jumping. Elevated arousal has been associated with poorer task performance (Bray *et al.*, 2015), emotional overwhelm, greater intensity in certain behaviours, and even the emergence of abnormal behaviours (Tooley and Heath, 2023). It is believed that heightened arousal can disrupt sleep, which is essential for emotional regulation and general well-being (Tooley and Heath, 2023). Caregivers often report that their dogs struggle to 'switch off', which may be a sign of chronic emotional dysregulation resulting from over-stimulation. Moreover, because balls are highly valued by many dogs, they can become sources of conflict during walks, sometimes triggering aggression in dogs who were previously tolerant of others.

In humans, excessive exercise has been shown to negatively impact mood and contribute to gut dysbiosis (Robinson *et al.*, 2021). Emerging research into the canine gut–brain axis suggests similar risks for dogs, highlighting how overexertion may negatively affect both physical health and emotional balance (Batson, 2024).

So, to fetch, or not to fetch?

So many dogs enjoy fetch and taking this enjoyable activity away from them completely is something we should hesitate to do. However, there are some low-impact, low-arousal alternatives that avoid many of the complications listed above. Try making these changes for safer play outdoors:

- Avoid using ball throwers (unless needed for accessibility reasons); instead, try gently rolling the ball along the ground.
- Stick to playing on flat, even terrain. Don't ask your dog to chase balls downhill or leap through the air.
- Incorporate impulse control into the game. If your dog is physically able, ask for a 'sit' and 'wait' before each throw to help prevent overexcitement.
- If a 'sit' isn't achievable yet, consider teaching a 'look at me' cue as an alternative way to pause and connect.

- Building on the previous point, once your dog is comfortable waiting, you can try hiding the ball and asking them to search for it, adding valuable mental stimulation!
- Limit fetch to three to five throws at a time, followed by a 10 min break. You can say 'all finished' and encourage your dog to explore their surroundings. They may continue to seek the ball for a bit, but it is worth persisting.

Alternatively, taking some treats and teaching a dog a game of 'find it' is a great option, particularly where there are dog friendly bushes or long grass to have them explore. Caregivers may also find that dogs become more social once they are not fixated on the ball, and so they may benefit from short periods of play with other friendly dogs.

Finally, dogs should not be dependent on a ball (or a game of fetch) for every walk. It can be tempting to lean on this popular activity and for every walk to be easy, but it is likely to be at the detriment of the dog.

Agoraphobia and Walk Refusal

For some dogs, the idea of going outdoors is too much. For others, certain locations may become problematic, but they may cope with walking in others. Either way, walk refusal (often similar to agoraphobia in humans) is very frustrating for caregivers.

There are many scenarios and motivations that can lead to walk refusal but the two most common emotions associated are fear and pain. These emotions may be caused by:

- The location.
- Associated car travel.
- Associated equipment (e.g. fear of the harness).
- A particular stimuli encountered on walks (e.g. noise fear after a dog has been exposed to bird scarers).

When starting treatment for this presenting complaint, there are two fundamental treatment strategies that must be implemented right away:

- Full health review from the referring veterinary professional.
- Walks become *optional*.

In most cases, this behaviour has come about because the dog does not feel safe. As seen throughout this book, in order to create safety there has to be choice, eventually leading to consensual activity. Dogs can be offered a walk, but should be allowed to refuse if they feel unable. There is no benefit to forcing a dog to go out when it is too frightened to cope.

In the meantime, caregivers can work with a Clinical Animal Behaviourist on a gradual programme of desensitization.

Building Confidence (Inspired by Free Work)

To encourage exploratory behaviour and increase confidence in the face of novelty, a simple activity taking inspiration from Sarah Fisher's 'Free Work' can be beneficial.

> As well as being an enriching experience on its own (and physical well-being is an important part of any enriching experience), ACE [Animal Centred Education] Free Work

helps to reduce stress, improve posture and balance, and increase the range of natural movement patterns. Free Work also helps to engage the seeking system, meet a dog's natural desire to be curious, and supports the dog's internal environment by helping them to release body tension, reset, and relax.

(Fisher, 2019)

Taking the idea of using novel stimuli and encouraging seeking behaviour, setting up an enrichment space for confidence building is highly effective in helping dogs to overcome fear. Caregivers simply need to fill the space with toys, boxes, bags, buckets (whatever is accessible) at varying heights and filled with plenty of treasure (Fig. 8.2).

In the context of dogs that fear the outdoors, this exercise is great to enrich the garden, driveway and eventually area around the front of the house in order to encourage them to step out of their comfort zone. Using a long line, a caregiver can simply follow their dog as they explore, each session stretching the boundaries as their confidence grows. Dogs will learn that there is reward in trying new things and learning to face their fears!

Muzzle Training

If a dog has ever bitten or shown intention to bite, they should be muzzle trained prior to practising any of the below behaviours in keeping with the law (Dangerous Dogs Act, 1991). Failing to prevent a bite or keep a dog under control can lead to a fine or even a prison sentence. Though absolutely necessary in the case of these dogs, it is never a bad idea to muzzle train a non-aggressive dog either, given that they may find themselves in circumstances beyond our control and require an additional safety net.

Fig. 8.2. Honey enjoys a course of novel stimuli inspired by Sarah Fisher's Free Work. Given that we are treating noise sensitivity, her caregivers have specifically added 'noisy' stimuli for her to explore. Image is author's own.

Choosing the right muzzle

An abundance of muzzles is available, but unfortunately many are uncomfortable, aversive or simply not fit for purpose. To ensure that a dog can take rewards, drink water and pant freely they should be fitted with a basket muzzle in the correct size. Many muzzle companies offer fitting advice or in-person fitting appointments (Fig. 8.3).

Introducing the muzzle

Unless deemed an absolute emergency, the introduction of the muzzle should not be rushed. If at all possible, in a situation where a dog is deemed unsafe to go outside without a muzzle they should have a few days at home to muzzle train rather than deal with the shock of being muzzled before they are ready.

Allow time to explore

A great way to introduce a muzzle is to place it somewhere as part of a game or 'Free Work' type activity. A simple way of doing this is to leave it on the floor (or on a surface) with a few treats inside and around the edges, allowing the dog to explore at their own speed. This is a particularly important step if the dog already has a negative association to the muzzle (for example they have been hastily muzzled in an emergency at the vets). There should be no pressure for the muzzle to go onto their nose at this stage (Fig. 8.4).

Fig. 8.3. A well-fitting basket muzzle demonstrated by @neo_therescuepup. Image is author's own.

Fig. 8.4. Neo explores the muzzle. Image is author's own.

Encourage a voluntary nose into muzzle

The next step is to encourage a nose into the muzzle, and this should only happen when the dog is engaged and showing that they are confident about what happens next (tip: if they are still nervous, try this next step with a cup or flower pot first).

Hiding a treat or some paste in the bottom of the muzzle should encourage the dog to put their nose in for the reward. After multiple repetitions a cue can be added such as 'nose' (Fig. 8.5). From here, treats can be continually fed to encourage duration and to desensitize the dog to reaching right into the muzzle.

Introduce the straps

Next, caregivers can introduce the straps. This step should be built up incrementally, as it is easy to move too fast and spook the dog. First, bring the straps around the head before immediately undoing them again, introducing the sensation to the dog but without lingering long enough for them to feel trapped (Fig. 8.6).

Slowly increase duration

Once the dog is suitably desensitized to the straps being applied the caregiver can start to build duration. This is done most effectively by short bursts with lots of rewards (Fig. 8.7). From here, it is sensible to gently expose your muzzled dog to stimuli around them, ensuring that they do not completely freak out when removed from a training setting in their home.

Fig. 8.5. Neo demonstrates a 'nose' cue into the muzzle. Image is author's own.

Fig. 8.6. The straps are clipped, and then immediately unclipped. This is repeated multiple times. Image is author's own.

Proofing the muzzle

Muzzled dogs should not be exposed to their triggers before they are comfortable with their muzzle. Otherwise, the muzzle will probably increase their emotional arousal and lead to poor decision making. Once they are ready, it is sensible to expose the dog from a great distance and increase proximity a tiny bit at a time.

Fig. 8.7. Reviewing some basic training with a high rate of reward is a great way to boost confidence for a newly muzzled dog. Image is author's own.

It is also good practice to use the muzzle outside of this setting, particularly if it can be paired with something highly rewarding. Failure to do this may cause the muzzle to become a predictor of negative stimuli or situations. To give an example of how to avoid this, if a dog is frightened of people but loves to be around dogs, it is worth muzzling them for short periods when playing with other dogs in order to build positive associations.

Appropriate Greetings

A lot of problem behaviour can be avoided if caregivers know which greetings are appropriate, and which to avoid. For example, when working with a dog who reacts out of fear to other dogs, it is completely counter-productive to orchestrate an interaction with a bolshy dog that invades their space.

Similarly, when trying to create positive associations with people, caregivers should meticulously select for the type of people that they wish to entrust their dog with. One over-the-top greeting is enough to undo months of hard work if the greeter is chosen incorrectly, or is poorly advised.

This section will provide the basic set-ups for interactions with dogs and people. An important thing to remember is that these guidelines are not set up for human convenience, but rather for optimum success and emotional wellbeing. Not only is it important to set up interactions as they are described, but it is equally important (if not more so) to *actively avoid* all other situations. Helping a dog to feel safe outdoors is first and foremost about forgetting societal expectations and being an advocate for them.

Unfamiliar dogs within a home

The following information will help a caregiver to perform an introduction between their dog and an unfamiliar (but appropriate) dog while outdoors. Introductions between dogs expected to live together follow a different protocol and should be performed slowly over a period of time. *The Art of Introducing Dogs* by Louise Ginman (2013) provides a detailed explanation of how to do this.

Unfamiliar dogs outdoors

First, dog-to-dog greetings should not be performed without consent from both human parties. As advocates for our dogs, we have a responsibility to say 'no' to inappropriate interactions that may harm their wellbeing, and we also have a responsibility to afford others the same opportunity to decline. For this reason, this protocol is best performed with known people (and their unknown dogs) that can share any observations, comments or concerns. Where enthusiastic consent cannot be given (for example, the caregiver is too far away), it is not appropriate to try to introduce a dog or simply to 'risk it'.

Second, not all dogs want to meet others. If the individual in question chooses to stay away from other dogs, demonstrates avoidance and reacts aggressively when cornered, it is worth considering whether their preference would be to avoid dogs altogether. In this instance, humans should be respectful, since continually forcing a dog into a situation they do not enjoy is a welfare and wellbeing compromise.

Preparation

In order to perform an introduction between unfamiliar dogs both canine parties should be at a sensible level of arousal to begin with. This can be established by maintaining consistent engagement with a handler as well as not reacting explosively at the sight of the other. For this reason, greetings often benefit from distance work beforehand.

A brief reaction followed by fast recovery is not necessarily a reason to abandon any of the below exercises, but if a dog is experiencing moderate to severe difficulty with any of the following steps their handler should withdraw and move more slowly through the process.

Reward calmness from a distance

The simplest part of this process is to just exist in the presence of the unfamiliar dog. This may mean walking at the opposite end of a field, or it may mean sitting on a park bench and watching the other dog engage with their handler. Throughout this exercise, all calm and otherwise desirable behaviour should be rewarded. This indicates to the dog the kind of mindset that is appropriate around the other dog, and also makes the other dog a predictor of good things.

Follow the leader

Once calmness is heavily reinforced, the dogs can start to walk in closer proximity. Following one after the other with ample distance between is a non-confrontational way for dogs to start to gather information about one another.

It may become apparent that one dog is not happy being followed by the other, or that one dog is struggling being at the back. In these situations, it is preferable to change direction to put the other dog at the front, as causing frustration is counter productive.

Parallel walking

An extension on 'follow the leader', parallel walking requires slightly more proximity, but still does not require dogs to be within touching distance. This can be done 20 metres apart and still be effective; it can also be done either side of a fence if neither dog is frustrated by barriers.

Three-second greeting

It is frequently said that on-lead greetings between dogs lead to reactive behaviour or dog fights. There is no evidence to support this claim, and in fact in a survey of 206 caregivers questioned after a dog-bite incident, 56.3% shared that both dogs were off-lead at the time of the incident (Roll and Unshelm, 1997). It is certainly likely that dogs feeling trapped or becoming tangled in leads would be more likely to direct aggression towards one another, but a quick and controlled on-lead greeting is no more risky than any other method.

Preferably, dogs should orient closer in the shape of a curve as shown in Fig. 8.8, mimicking natural dog to dog greetings. Handlers can allow their dogs to greet for a count of three, and then excitedly cue them to move apart (for example, 'let's go'). This brief interaction is in line with organic dog greetings, lasting between 0.8 and 6 s (Ward, 2020). Failing to move away within this window can cause two on-lead dogs to become worried or frustrated and make undesirable behaviour choices.

If this interaction has gone well, handlers may choose to go onto the next step. If the dogs still seem measurably tense, or if one chooses to react, they will benefit from some distance work while recovering before attempting another interaction.

Off lead interaction (if appropriate)

Off-lead interaction may follow some time on-lead. In this instance mutual calmness, or calmness with intermittent play is desirable. To keep this interaction successful:

- Ensure that one dog is not harassing the other; the pushier dog should be interrupted if it continually pursues the other and ignores their body language.
- Avoid the introduction of high-value resources if possible; balls and treats can lead to conflict.

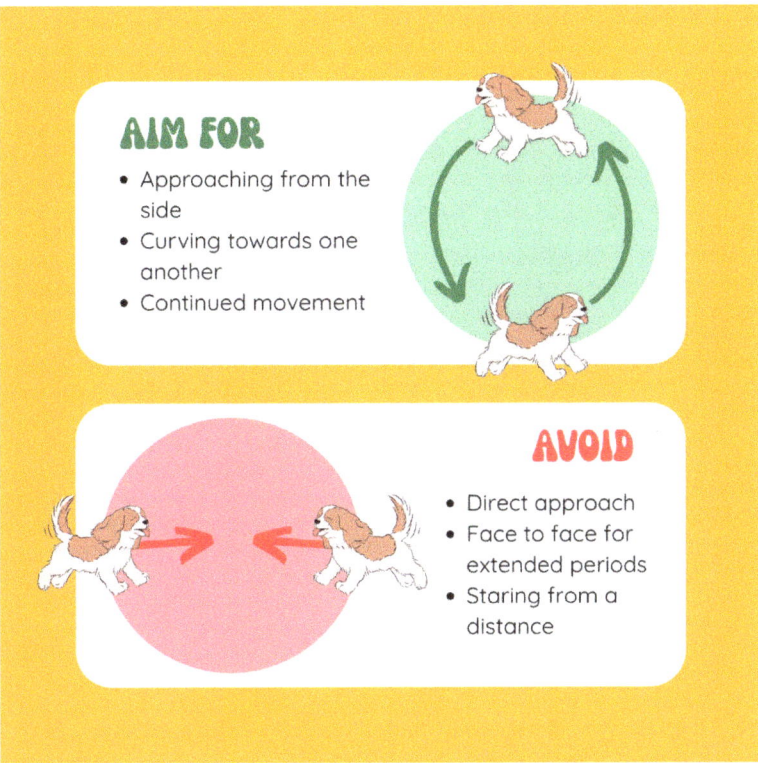

Fig. 8.8. Dogs should approach one another in a curving motion, avoiding head-on greetings. Image is author's own.

- Encourage sniffing between short bursts of play; long periods of play lead to high arousal and bad decisions!
- Don't panic if there are minor disputes that do not lead to a fight; dogs are allowed to communicate!

People

This section will be prefaced by a similar message to the sentiment about unknown dogs. Dogs do not owe humans anything, and so they should not be forced to interact with an unknown person if they do not want to. This is particularly important for those that generate a lot of attention when outdoors. Handlers must have the confidence to say 'no', and the general public must understand that they are not owed an explanation. If a dog would prefer not to be approached by strangers we should support them in that decision.

Having said that, if an introduction to a friend or family member needs to be set up with a dog that is worried about people there are ways to do so.

Muzzle train

As previously mentioned, if a dog has a bite history or is considered a bite risk, they should be muzzle trained prior to an interaction and should be wearing their muzzle to greet.

Human body language

Instructions must be given to the unfamiliar human in question beforehand to ensure that they behave appropriately. The advice given below is based on decades of observation of canine body language, and mimics how unfamiliar dogs choose to greet one another.

Behaviours to avoid:

- Being loud or excitable.
- Leaning or reaching over a dog's head.
- Sharp or unpredictable movements.
- Physically restraining the dog (for example in a hug).
- Crouching down to greet (from a safety perspective this should be avoided).

Instead, do this:

- Remain calm and still.
- Hands should stay at the side.
- If the dog becomes worried or confrontational, the best thing to do is turn away.
- Walk alongside the dog for a while and quietly talk to it in a 'sing-song' tone.

Appeasement signals

All humans should be familiar with appeasement signals, and they should be confident in their ability to move away and diffuse a situation should the dog require it. Most commonly, dogs will look away and/or lick their lips if uncomfortable with human interaction (Firnkes *et al.*, 2017). If these signals are present, it is a good idea to take a break from the interaction and allow the dog to decide whether to proceed or not. Some dogs may settle quickly, whereas others may need multiple repetitions of the above guidance before being able to cope with the individual.

A Note on 'Doggy Day-Care' and Dog Walkers

Using either a dog walker or 'doggy day-care' service is very popular, which is hardly surprising given that caregivers are rarely fortunate enough not to need full-time jobs. In choosing to do so, there are risks ranging from minor (slips and trips) to major (dog fights and lost dogs), as well as the risk of an individual developing behavioural problems.

There are, however, benefits to using both services, but it is incredibly important that caregivers are informed for the sake of the safety and wellbeing of their dog. Most notably, if a caregiver is working hard with a behaviour professional but their chosen service is undoing all of their hard work through bad practice, their valuable time, money and patience will ultimately be wasted.

Doggy day-care

In order to run a day-care facility in the UK a business must have a licence. The 'Dog day care licensing: Statutory guidance for local authorities' (Department for Environment,

Food & Rural Affairs, 2025) provides rough guidance for how to run such a premises, but it can be difficult for even these minimal regulations to be enforced. For example, the minimum standards state that 'each member of staff should have 10 dogs or less to care for', a ratio that would make most canine care professionals recoil in horror. If a fight were to break out between 10 dogs and one member of staff the outcome does not leave much to the imagination.

This is not to say that there is not a place for day-care facilities. In fact, providing this valuable service prevents dogs from being left alone for an 8 hour workday (or longer), and even prevents relinquishment due to unmet needs. The purpose of this discussion is not to dissuade the reader from using day-care facilities altogether, but rather to ask them to be scrupulous when deciding who to trust with their dog.

There is little research concerning the safety, welfare and behaviour of dogs at day-care facilities, but based on other research in the canine field as well as first-hand observational experience, the risks are as follows:

- Dog fights, particularly in large groups with few staff.
- Repeated impact of play and running on hard floor, possibly leading to musculoskeletal problems.
- Consistently high arousal (refer to 'the dangers of fetch' above).
- Poor matching between dogs leading to inappropriate development of play or unwanted social learning of habits.
- Poor matching leading to dogs becoming fearful (often pursued by another dog without a break).
- Dogs prone to injuries being unable to rest.
- Contagious diseases such as kennel cough.

For this reason, caregivers should consider the following before choosing a facility:

- Request the health and safety policy as well as the emergency dog bite procedure.
- What vet are they registered with? Are you happy with the reputation of that vet?
- Are they qualified? Will they be knowledgeable about dog behaviour?
- What methods do they use to communicate with their dogs? Are they going to be punitive?
- How many members of staff per dog are on site?
- How do they ensure dogs are resting throughout the day? (particularly important for dogs under 18 months).
- Is the floor slippery or hard? If so, how do they reduce the possibility of injury?
- Are all dogs vaccinated (particularly against kennel cough)?

Additionally, on a personal note, I feel that the staff should easily be able to communicate with customers throughout the day. This helps to alleviate any caregiver concerns as well as ensuring that messages are delivered smoothly in emergencies.

Dog walkers

Dogs are typically with a dog walker for a shorter duration than they would be at a day-care facility, so caregivers have a bit more control over rest periods and decompression after all that excitement! However, the main points regarding health and safety, veterinary access and behaviour knowledge still apply.

Additionally, it is worth finding out in greater detail what methods they use to handle their dogs. Given that dog walkers are often intermingling with the general public, they are more likely to experience unpredictable situations. They should know how to deal with a variety of incidents calmly and safely without needing to get frustrated or be heavy-handed. They should also be trained in Canine First Aid.

Coping with Reactivity

> Your dog is not giving you a hard time. **Your dog is having a hard time.** – Unknown

While this next section is designed to help those experiencing problems with reactivity, many of the exercises are incredibly *helpful for all dogs to learn*. Focusing on the dog–caregiver dynamic and increasing engagement makes a difference to all dog walks.

What is 'reactivity'?

It is not entirely clear how prevalent reactivity is in the UK dog demographic, possibly because there is not a nationally understood definition. The 2024 PDSA PAW report saw 11% of respondents sharing that their dog had growled, snapped at or bitten an unfamiliar dog (PDSA, 2024). In the National Dog Survey, 17% of respondents shared that their dog had difficulty staying calm around other dogs (Dogs Trust, 2024). These results likely represent something close to the prevalence of what we know as reactivity in the discussion of dog behaviour.

'Reactivity' is not a diagnosis. Describing a dog as 'reactive' is simply describing what the dog is doing, but is not attempting to explain the underlying cause.

'The word 'reactivity' translates dogs' responses to different stimuli, especially with regard to intensity and length of the response, the time of the reaction, and the time needed for the dog to regain homeostasis' (Sforzini *et al.*, 2009).

To continue, describing a dog as reactive does not automatically mean that they are exhibiting aggression or intention to cause damage. Dogs may react through frustration, desire (wanting to play) or simply through habit. They may direct their behaviours towards dogs, people, traffic or something else entirely, and their motivation when reacting to these stimuli may vary. To describe as a dog as performing 'reactive' behaviour simply indicates that they are reacting to something in the environment.

All this said, the basic skills and exercises for treating all presentations of reactive behaviour are often the same. Firstly, caregivers need to have a secure relationship with their dog when outside, and this can be achieved by increasing engagement and ensuring that both parties are enjoying one another's company.

Next, dogs need to go through a process of systematic desensitization and counter conditioning (more on this below) in order to teach their dog alternative behaviours and change the way that they view the situation. With these two processes in place, caregivers can expect to be able to manage their dog happily from a distance and ultimately find a lot more peace on their walks. If they wish to progress further after this time, a Clinical Animal Behaviourist is required to diagnose the root cause of the reactive behaviour and treat accordingly.

Adjusting your mindset

Stressed humans create stressed dogs, and heavy restrictions create frustration and dogs that lash out. While it is very important to select the right equipment for a dog (particularly one with behavioural difficulties), it is equally important to manage how we use it.

Safety comes first. But, when working at a reasonable distance from a trigger it is preferable to keep a loose and relaxed leash and to aim for calm cues and body language. Jerking the lead or insisting that a dog walks to heel is a sure way to make them feel more uncomfortable and frustrated. Similarly, constant micro-management (particularly if reprimanding) is completely counterproductive to the low stress environment we are trying to create. Instead, caregivers should learn to relax their body and breathe slowly, allowing the dog to process information without constant input.

Exercises for engagement

The following are all staple exercises for increasing engagement with a dog outdoors. They will likely need to be practised somewhere quiet to begin with before being introduced to triggers at a distance. With time and practice, they should become automatic, and be used in close proximity to triggers or in emergencies.

'Check-ins'

'Check-ins' are a simple response to name that can be used whenever a dog has started to disengage on a walk (Fig. 8.9).

1. Cue the dog's name.
2. As the dog looks towards you mark (either using a clicker or a verbal marker such as 'yes') and reward.
3. Repeat whenever the dog disengages for a period of time.
4. Over time, dogs will learn that it is valuable to check in and will start to offer it as a behaviour.

'This way'

'This way' can be used to change direction if a dog is fixating, or if a trigger is oncoming and their handler wishes to quickly move away (Fig. 8.10). By developing a strong 'this way' cue that is as close to automatic as possible, handlers can avoid or escape from situations quickly.

1. Cue the dog's name (as per a 'check-in').
2. Stop, and gently press on the lead to turn the dog around (they should turn their back to the stimuli).
3. Cue 'this way' and walk backwards with energy, rewarding as the dog follows (Fig. 8.11).
4. Resume walking as normal in the new direction of travel.

Fig. 8.9. Cheddar offers a stationary check-in. Image is author's own.

Fig. 8.10. Cheddar moves ahead in front, so I stop, ready to turn him around. Image is author's own.

Fig. 8.11. I gently encourage him to turn around by pressing on the lead and walking backwards. I cue him 'this way', and reward him as he follows me. Image is author's own.

Desensitization and counter-conditioning

One of the key components of any behaviour modification plan is likely to be a desensitization and counter-conditioning protocol. Desensitization is the process of gradually and repeatedly exposing an animal to a stimulus at a level that they can cope with, over time seeking to reduce their emotional response. Counter-conditioning teaches an alternative emotional response to the stimulus, over time changing the thought. These techniques are often used alongside one another, but dogs with a particularly adverse emotional response to stimuli may require a process of systematic desensitization prior to starting counter-conditioning.

Watch the world

Desensitization can be done easily with minimal effort. For many individuals, this will be as simple as sitting and watching the world go by. An example of this method in practice is used when working with dogs experiencing agoraphobia or walk refusal. By simply sitting on the doorstep and allowing the dog in question to 'watch the world' they can become desensitized to life outdoors without being put in a situation they cannot cope with. This is also very similar to puppy socialization when done correctly!

Engage/disengage

For dogs that may struggle to sit and watch a trigger without becoming distressed or over-stimulated, an 'engage/disengage' exercise gives them something to think about. This

Fig. 8.12. Allowing Rufus to watch the stimuli (engage) before asking for him to look back to handler (disengage). Image is author's own.

exercise is not intended to stop them from looking at a stimulus, but rather to interrupt them and prevent them from fixating. With repetition and consistency, it can become an automated response for dogs that are worried by a particular stimulus.

It looks like this:

- The dog looks at a stimulus (they should be far enough away from it to remain calm) (Fig. 8.12).
- Put a treat under the dogs' nose and bring it up to you, orienting them to look towards you.
- Mark the behaviour and release the reward (Fig. 8.13).
- With repetition you will be able to phase out luring with the treat and add a voice cue such as 'look at me'.
- Over time, you will need to prompt the dog less and less.

To walk, or not to walk?

For a long time, we were told that all dogs need two walks a day, every day. For many dogs, particularly those dealing with complex behaviour issues, this is simply not the case. If a caregiver, or their dog (or both) are not going to be the best version of themselves on a walk, they should simply do something else. Knowing that emotional arousal and trigger stacking impacts the decision making of our dogs, it is sometimes in our best interest to take a day or two for recovery. The same applies for humans, and if a caregiver is not in the headspace to support their reactive dog through a walk that day, it is unlikely that their walk will be constructive or positive (more about this below).

Fig. 8.13. Rewarding for the disengage from stimulus. Image is author's own.

Alternative activities to entertain dogs include enrichment, training sessions and even a 'garden safari'. This is created by gathering pieces of the outside world (normally plant debris) and decorating the dog's own garden, bringing important information from the outside into the safety of their familiar surroundings.

A message to caregivers

Behaviour problems can make life very lonely. Arguably, when problems are occurring within the home they can be managed privately. However, when dogs become reactive or otherwise problematic outside it can be embarrassing, scary and sometimes unsafe. Speaking from experience as a practitioner, *nobody is alone in this battle*, and *almost everyone feels the same*. A supportive group, walking companion or online friend can be a great way to boost confidence and let off steam on the bad days. Having a dog with behaviour problems is nothing to be ashamed about. It is how you proceed that matters.

Car Travel

Problem behaviours related to car travel are very common, though the emotional response varies. Some dogs find car travel incredibly stressful due to over-excitement and desire to get to where they are going. Others are terrified, possibly from an aversive experience or possibly due to memories that go back as far as they can remember. Some dogs even experience mixed emotions and find car travel a frustrating and emotionally conflicting situation.

As with most other behaviour problems, there is no quick fix for dealing with aversion to the car. From experience, the first thing is to establish that there is no pain or discomfort impacting the behaviour since both the physical act of jumping into the car and the experience of being restricted in a tight space for a period of time are horrible for painful dogs. Once the physical comfort of the dog has been optimized, the caregiver is required to start changing the emotional response to the car itself.

Reducing travel

If a dog is required to undertake training or behaviour modification around the car, they need to be allowed time for decompression. This means that car travel every day is counter-productive and should be reduced to one or two training journeys a week, giving them adequate time to recover from their last journey.

Some dogs with extreme phobias will benefit from stopping car travel altogether, making contact with the car only during consent-based games that they *choose* to partake in and are able to come away from at their preference. As their aversion reduces, they can be reintroduced slowly to physically travelling in the car. While this may make life difficult for caregivers temporarily it is often a necessary sacrifice for a temporary period of time.

Safety signals

While a dog is partaking in behaviour modification around the car, it is recommended that interactions remain consent based. This means that the dog chooses how much to engage in the training (if at all) and can also choose when to stop. As seen in other situations (Chapter 9, Co-Operative Care) dogs are more likely to engage in an activity if they feel that they have control over it.

There may come a point, however, where a dog needs to be transported against their will (for example, for an emergency vet appointment), before they are emotionally ready to do so. To avoid breaking the trust formed through consent-based training, 'safety' and 'non-safety' signals are introduced. These two signals are created using a strong environmental signal that differentiates between two situations.

Safety signal = A stimulus that indicates to the dog that they are in a safe, consent-based training environment.

Non-safety signal = A stimulus that indicates to the dog that this is a mandatory 'one-off' situation that they do not have a choice in.

For example, a caregiver may choose to do game-based training around the car using a harness and long line (safety signal) when the dog is in control, but use a collar and short lead (non-safety signal) to take them to the vet. These differences create confidence in the predictability of the situation and avoid undoing the hard work put into everyday training.

Comfortable access

Regardless of whether pain or discomfort is present, a dog should be able to access the car easily to avoid unnecessary stress. To facilitate this:

- Experiment between different areas in the car to see if one location is preferable (for example, some dogs find it easier to get onto the back seat than into the boot).
- Consider providing steps or a ramp.
- Provide a comfortable mattress or bed, particularly if a dog is jumping into the car.
- Park the car somewhere easily accessible, slopes and steps are enough to put some of them off.

Some dogs may prefer to have a cue word to enable a caregiver to lift them, but this should be taught as a co-operative behaviour. For example, teaching 'paws up' onto the lip of the boot enables the dog to signal that he is ready to go and the caregiver can help him up. This technique should not be used if the dog shrinks away or starts to panic when a human is too close.

Games for the car

Enrichment

For dogs with less severe responses to the car, simply practising calm behaviour may be enough to change their emotional response. With the car stationary, caregivers can spend a short period each day providing enrichment or having a cuddle, gradually leading up to doing so with the engine on and moving towards a desensitization schedule in transit (Fig. 8.14).

'Find it'

Some dogs are so emotionally triggered by the car that their behaviour will change at the slightest indication of car travel. For some dogs, this may involve refusing to move past a certain point, for others it may involve hiding as soon as their caregiver picks up the keys. In these sorts of cases a much more gradual approach is required, typically starting with a fun scent-based game and *no travel*.

First, the dog needs to be taught the game of 'find it' in a setting completely separate from the car. Usually, the garden is best. To teach the game:

- Choose a starting position and put the dog into a 'sit' and 'stay' (often it is easiest to provide a clear visual cue such as a towel on the floor).
- Hide a selection of treats (or a toy if this is more of a motivator).
- Return to the dog and cue 'find it'.
- Allow the dog to search the environment, only guiding them gently if absolutely necessary.
- Once they have cleared the treats or retrieved the item bring them back to the starting position and cue 'all finished'.

This should be repeated regularly until the dog is completely fluent and shows no hesitation completing the task. From this point, the game can gradually be moved closer and closer to the car, first encouraging investigation of the periphery before eventually braving the interior of the car. If at any stage the dog does not want to go past a certain area, the

Fig. 8.14. Change the association with a vehicle by providing enrichment and somewhere comfortable to sit. Image is author's own.

game should be finished and the treats put away. The area they find difficult can then be repeated in future sessions until they begin to feel more comfortable.

Desensitization

At the point that a dog is jumping into the car calmly, or with very little hesitation, desensitization to car travel can start. Each session should use a continuous schedule of reinforcement to keep motivation high and should be practiced multiple times before moving forward. A program of systematic desensitization over a period of weeks, or possibly months, will look something like this:

- Engine on, door/boot open.
- Engine on, door/boot closed, car stationary.
- Car moving back and forth on the driveway.
- Car driving a short distance and building slowly each time.

Additional measures

Some dogs will need additional support in order to overcome their aversion to car travel. Vets may support the use of behaviour medication in some cases (when used alongside professional behaviour modification) (Flint *et al.*, 2024). Research also suggests that CBD oil can be effective in reducing stress caused by car travel (Hunt *et al.*, 2023; Flint *et al.*,

2024) and pheromone collars may be effective when used to reduce car associated problem behaviour (Gandia Estellés and Mills, 2006). *Both of these treatment options should be discussed with a veterinary professional prior to use.*

Conclusion

Venturing outside can be highly stressful for both dogs and caregivers alike. If a dog feels unsafe or frustrated, they can develop all manner of behaviour problems that make walking difficult, and so it is important to work with a Clinical Animal Behaviourist on the emotional response to their triggers. Overall, it is most beneficial to introduce a consent-based approach to walks and ensure that the dog feels safe and supported. This can be done by being selective about the environment, being a calm and capable caregiver and allowing the dog time off to decompress when they need it.

Finally, it is important for all caregivers to remember that while our dog's emotional well-being is of the utmost importance, it cannot be at the detriment of public safety. Therefore, all possible measures should be taken to keep both parties safe. If caregivers are unfamiliar with their legal responsibilities, they should refer to the Dangerous Dogs Act (1991).

Notes

[1] Ethograms have shown that the four behaviours orient, stalk. chase and grab/bite have significant variation across breed groups, with the largest differences seen at the search and approaching parts of the sequence (Broseghini *et al.*, 2024).

References

Animal Welfare Act, A.W. (2006) Available at. Available at: https://www.legislation.gov.uk/ukpga/2006/45/contents (accessed 15 July 2025).

Baldursdóttir, G.H. (2024) Nosework and the Dog-Human Relationship. Msc thesis, University of South Eastern Norway, Norway. Available at: https://openarchive.usn.no/usn-xmlui/handle/11250/3172227 (accessed 15 July 2025).

Barn Hunt International (2025) About Barn Hunt UK. Available at: https://barnhuntuk.co.uk/about/ (accessed 15 July 2025).

Batson, A. (2024) Getting to the guts of dog behaviour. The Dog Genius. Available at: https://www.thedogenius.com/course/getting-to-the-guts-of-behaviour-with-dr-amber-batson (accessed 15 July 2025).

Bray, E.E., MacLean, E.L. and Hare, B.A. (2015) Increasing arousal enhances inhibitory control in calm but not excitable dogs. *Animal Cognition* 18(6), 1317–1329.

Broseghini, A., Lõoke, M., Guérineau, C., Marinelli, L. and Mongillo, P. (2024) Ethogram of the predatory sequence of dogs (*Canis familiaris*). *Applied Animal Behaviour Science* 279, 106402. DOI: 10.1016/j.applanim.2024.106402.

Dangerous Dogs Act, D.D. (1991). Available at: https://www.legislation.gov.uk/ukpga/1991/65/contents (accessed 11 July 2025).

Delgado, M.M., Stella, J.L., Croney, C.C. and Serpell, J.A. (2024) Making fetch happen: Prevalence and characteristics of fetching behavior in owned domestic cats (*Felis catus*) and dogs (*Canis familiaris*). *PLOS One* 19(9), e0309068.

Department for Environment, Food & Rural Affairs (2025) Dog day care licensing: Statutory guidance for local authorities. Available at: https://www.gov.uk/government/publications/animal-activities-licensing-guidance-for-local-authorities/dog-day-care-licensing-statutory-guidance-for-local-authorities (accessed 15 July 2025).

Dogs Trust (2024) Welcome to the Results of the National Dog Survey 2024. Available at: https://www.dogstrust.org.uk/downloads/Dogs_Trust_NDS_Report_2024__.pdf (accessed 10 July 2025).

Firnkes, A., Bartels, A., Bidoli, E. and Erhard, M. (2017) Appeasement signals used by dogs during dog–human communication. *Journal of Veterinary Behavior: Clinical Applications and Research* 19, 35–44. DOI: 10.1016/j.jveb.2016.12.012.

Fisher, S. (2019) Animal Centred Education. Available at: http://www.tilleyfarm.org.uk/AceIndex.php (accessed 15 July 2025).

Flint, H.E., Hunt, A.B.G., Logan, D.W. and King, T. (2024) Daily dosing of cannabidiol (CBD) demonstrates a positive effect on measures of stress in dogs during repeated exposure to car travel. *Journal of Animal Science* 102, skad414. DOI: 10.1093/jas/skad414.

Gandia Estellés, M. and Mills, D.S. (2006) Signs of travel-related problems in dogs and their response to treatment with dog appeasing pheromone. *Vet Record* 159(5), 143–148.

Ginman, L. (2013) *The Art of Introducing Dogs*. Balboa Press, Australia.

Hunt, A.B.G., Flint, H.E., Logan, D.W. and King, T. (2023) A single dose of cannabidiol (CBD) positively influences measures of stress in dogs during separation and car travel. *Frontiers in Veterinary Science* 10, 1112604.

Kokocińska-Kusiak, A., Woszczyło, M., Zybala, M., Maciocha, J., Barłowska, K. *et al.* (2021) Canine olfaction: Physiology, behavior, and possibilities for practical applications. *Animals* 11(8), 2463. DOI: 10.3390/ani11082463.

Lowe, B. (1981) *Hunting the Clean Boot*. Blandford Press, Poole, UK.

Marinelli, L., Adamelli, S., Normando, S. and Bono, G. (2007) Quality of life of the pet dog: Influence of owner and dog's characteristics. *Applied Animal Behaviour Science* 108(1–2), 143–156. DOI: 10.1016/j.applanim.2006.11.018.

Mason, G.J. and Burn, C.C. (2011) Behavioural restriction. In: Appelby, M.C. (ed.) *Animal Welfare*, 2nd edn. CABI, Wallingford, UK, pp. 98–119.

Mills, D.S., Demontigny-Bédard, I., Gruen, M., Klinck, M.P., McPeake, K.J. *et al.* (2020) Pain and problem behavior in cats and dogs. *Animals* 10(2), 318. DOI: 10.3390/ani10020318.

Mills, D.S., Coutts, F.M. and McPeake, K.J. (2024) Behavior problems associated with pain and paresthesia. *Veterinary Clinics of North America - Small Animal Practice* 54, 55–69. DOI: 10.1016/j.cvsm.2023.08.007.

Nganvongpanit, K., Tanvisut, S., Yano, T. and Kongtawelert, P. (2014) Effect of swimming on clinical functional parameters and serum biomarkers in healthy and osteoarthritic dogs. *ISRN Veterinary Science* 1–8. DOI: 10.1155/2014/459809.

Nganvongpanit, K., Ruamrungsri, N., Tepsoontorn, B., Yano, T., Siengdee, P. *et al.* (2016) Effects of swimming frequency on body weight and serum lipid profile in small-breed dogs during a four-month period. *Thai Journal of Veterinary Medicine* 46(4), 655–661. DOI: 10.56808/2985-1130.2785.

Paddock, A. (2023) How Many Types of Retrievers Are There? Get to Know All Six Breeds. American Kennel Club. Available at: https://www.akc.org/expert-advice/dog-breeds/retriever-breeds/ (accessed 15 July 2025).

PDSA (2024) PDSA Animal Wellbeing (PAW) Report. Available at: https://www.pdsa.org.uk/what-we-do/pdsa-animal-wellbeing-report/paw-report-2024 (accessed 10 July 2025).

Robinson, E., Thornton, E., Templeman, J.R., Croney, C.C., Niel, L. *et al.* (2021) Changes in behaviour and voluntary physical activity exhibited by sled dogs throughout incremental exercise conditioning and intermittent rest days. *Animals* 11(18), 118.

Roll, A. and Unshelm, J. (1997) Aggressive conflicts amongst dogs and factors affecting them. *Applied Animal Behaviour Science* 52, 229–242.

Rooney, N.J. and Bradshaw, J.W.S. (2003) Links between play and dominance and attachment dimensions of dog-human relationships. *Journal of Applied Animal Welfare Science* 6(2), 67–94.

Scent Work Academy (N.d) Scent Work Training with Your Companion Dog. Available at: https://www.scentworkacademy.co.uk/ (accessed 22 April 2025).

Sforzini, E., Michelazzi, M., Spada, E., Ricci, C., Carenzi, C. *et al.* (2009) Evaluation of young and adult dogs' reactivity. *Journal of Veterinary Behavior: Clinical Applications and Research* 4(1), 3–10. DOI: 10.1016/j.jveb.2008.09.035.

Sharkey, L. (2025) Are You Playing Too Much Fetch? https://www.kinship.com/uk/dog-behaviour/too-much-fetch-dog

The Border Collie Trainer. (2023) Sheepballs. Available at: https://www.thebordercollietrainer.org/lessons/sheepballs-info (accessed 15 July 2025).

Tooley, C. and Heath, S.E. (2023) Emotional arousal impacts physical health in dogs: A review of factors influencing arousal, with exemplary case and framework. *Animals* 13(3), 465. DOI: 10.3390/ani13030465.

Ward, C. (2020) Greeting behavior between dogs in a dog park. *Pet Behaviour Science* 10, 1–14. DOI: 10.21071/pbs.vi10.12314.

9 The Link Between Physical and Emotional Wellbeing

JADE NICHOLAS[1]* AND FABIAN RIVERS[2]

[1]CAB, Winchester, UK; [2]MVDR, MRCVS GPCert(ExAP), Birmingham, UK

Abstract

Pain and/or discomfort is expected to be present in approximately 80% of dogs presenting with behavioural conditions. Behavioural changes such as increased fear, separation problems and reactivity are thought to be early indicators of pain and discomfort, leading to a major welfare concern for dogs when these things are ignored. First, Fabian Rivers introduces the UK dog demographic through a veterinary lens. Then, this chapter explores behavioural signs of pain and discomfort including indicators of cognitive issues, gastrointestinal problems and musculoskeletal ailments. Once symptoms have been established, recommendations are given for low-stress examination as well as low-stress protocols for other stressful environments (such as grooming). Finally, an introduction to 'co-operative care' is provided for readers to begin teaching the dogs in their care. In order to achieve optimum emotional wellbeing, the physical side of canine welfare cannot be ignored.

Introduction

The consideration of pain and discomfort in the emotional wellbeing of our animals is unavoidable. Pet dogs are often seen by 'professionals' for training and behaviour concerns but are not assessed for pain and discomfort, leaving the underlying cause of the issue partially untreated.

The International Association for the Study of Pain (IASP) defines pain as 'an unpleasant sensory and emotional experience associated with, or resembling that associated with, actual or potential tissue damage' (IASP, 2024). The WSAVA *guidelines for the recognition, assessment and treatment of pain* go on to say that 'pain is not just about how it feels, but how it makes you feel' (Monteiro *et al.*, 2023). Since pain is unique to each individual, the guidelines express limitations caused by the subjective nature of the emotion. However, further guidance by the IASP reminds the reader that non-verbal patients may use behavioural indicators to tell us that they are in pain.

A study published by Mills *et al.*, in 2020, considered the widespread prevalence of pain in dogs displaying behavioural problems, revealing that somewhere between *30 and 80% of dogs referred for behaviour modification are likely to be experiencing pain*. Worryingly, the 2024 PDSA Paw Report revealed that not all dogs in the UK are currently registered with a vet (PDSA, 2024).

Before investigating the link between pain and behaviour problems, the following essay by Fabian Rivers will provide insight into the overall health of domestic dogs in the UK (2025).

*Corresponding author: aboutyourdog@outlook.com

'A Vet's Perspective' by Fabian Rivers

Pain, discomfort and sadness. Empathy from a vet's perspective

To practice veterinary medicine in the UK today is to sit at an odd and often frustrating crossroads. One foot is firmly in science, diagnostics and intervention – the one we normally are accustomed to. The other is in a deeply human domain of emotion, expectation and, increasingly, ethical concern. It is within this space that I meet dogs every day, each one absolutely a product not only of their biology, but of the cultural and economic environments we have built around them. In fact, the biology we have built for them as well.

And so many of them are in pain.

It is not always the profound, overt and dramatic pain of acute trauma or obvious injury, but something we know is more insidious. A quiet drum of discomfort that has been background noise to both caregiver and many veterinary practitioners, until it is simply considered part of life. The normal turning of the cogs of the animal space. I want to revisit the reflection on that discomfort. On how it presents, how it is and has become normalized, and how, in many cases, it is created by systems we have come to accept as 'normal'.

The evolving landscape of veterinary practice

Over the past decade, the work of a small-animal vet has shifted in subtle but significant ways. The sheer volume of cases has grown, shaped by increased pet 'caregivership' and the fragmentation of care between general practice, corporates, out-of-hours providers and social media advice. Meanwhile, expectations have intensified. Trust me, I know! Public trust is waning and, in many ways, the centres of disinformation about what is healthy, good practice and what is endangering and problematic is blurred and in many occasions inverted.

There are more French Bulldogs, Dachshunds and until recently XL Bullies (albeit not an accepted breed – more a vague class of animals) with designer names and complex needs. There are fewer insured clients, fewer routine check-ups, and a sharper divide between those who can afford pre-emptive care and those who arrive only when crisis has taken hold. It is a disaster in many ways for both animals and those who care.

Let me help cover a variety of issues which plague many dogs in the landscape of today.

Obesity

Obesity is one of the most prevalent preventable diseases in the UK dog population. Yes, it is in the right context considered not only a result of dietary indiscretion but also a disease, mainly because the side effects of weight have prevalent metabolically relevant effects beyond excess fat. Despite consistent messaging from the veterinary profession, its implications are still too often downplayed or misinterpreted. The consequences are substantial. Dogs carrying excess weight are at greater risk of osteoarthritis, exercise intolerance, heatstroke and diabetes. In many cases, they are also in pain, be it from a physical perspective or as mentioned due to secondary effects of obesity.

However, palpably in the clinic, these issues manifest their own difficulties. Behaviourally, pain-related changes may be subtle: hesitation when climbing stairs, increased rest periods, grumpiness or resistance when handled. These signs are often attributed to ageing rather than pathology. Sometimes there is some element of those concerns. However, The British Small Animal Veterinary Association (BSAVA) has consistently emphasized the need for early recognition and intervention, yet conversations around weight remain fraught with discomfort for both vets and owners (Hall *et al.*, 2020).

In truth, obesity-related pain is chronic, progressive and largely avoidable and/or manageable. As clinicians, we must remain committed to raising this issue, even when it risks tension. Compassionate honesty is key. Weight loss, where appropriate, is not simply aesthetic. It is a fundamental component of pain management and long-term health. Challenging and being proactive about obesity *and* its attached pain issues, not only helps better outcomes but also means we are not navigating confusion between pain or behavioural nuance, an issue which is highly difficult to address when so many dogs are obese.

Osteoarthritis: the unseen burden of movement

Osteoarthritis (known colloquially just as arthritis) affects as many as 40% of dogs, particularly in larger breeds and ageing individuals (Platt and Olby, 2013). Yet its behavioural expression is frequently misunderstood. Carers often report their dog is 'just getting old' and 'not interested in walks anymore,'. This is a common presentation. Or my favourite, 'doesn't seem to be in pain', when in reality, they may be avoiding pain-inducing activity.

Studies in behavioural medicine have drawn clear links between musculoskeletal pain and behavioural change. In one such study, 75% of dogs with impulsive aggression were found to be suffering from conditions such as hip dysplasia and elbow (osteo) arthritis (Camps *et al.*, 2012). These findings underscore the need for routine pain assessments in cases of sudden behaviour change. The central importance of taking pain in potentially arthritic dogs is of prime importance.

Personally, I have witnessed countless times that dogs labelled as irritable or unpredictable have later been diagnosed with chronic joint pain. I often have a litmus test, where pain is believed to be possibly arthritic and other concerns are not paramount to address, a short trial of pain relief can be an eye-opening experience. Navigating a low-risk intervention such as a non-steroidal anti-inflammatory drug (NSAID) for five to 7 days and then to observe behaviour during and after can provide a stark contrast to normalized behaviours. Once appropriate analgesia is initiated and maintained, countless cases I have witnessed have had dogs described as aggressive and fear ridden, transformed into softer demeanour patients. Their trust sometimes returns (not always with the vet admittedly). What was once considered a temperament issue is revealed to be a pain response. That is the power of pain.

Taking pain seriously is a *key element of happiness*. A happy dog is the top of the totem pole for the work I do daily.

Dermatological pain: the itch that speaks volumes

Dogs with allergic skin disease may not appear ill in the conventional sense, but the daily toll on their wellbeing is profound. Imagine the mosquito bite itch but all day, all night. Feet, groin, face, ears without reprieve. It affects sleep, comfort, social behaviour and emotional regulation. It is not simply an itch. It is distress. It is all encompassing and completely diminishes quality of life.

Veterinary dermatology texts note that chronic pruritus can lead to anxiety-like behaviours, social avoidance and frustration-based reactivity (Coatesworth, 2019). This reiterates a critical point: pain and discomfort are rarely isolated.

Too often, skin disease is managed symptomatically rather than systemically. Long-term management for issues that often are transient but long term should be valued more. Choices of drugs from a veterinary perspective should be aimed to build a clinical picture not only backed by science but a gameplan to acknowledge the long-term issues that are often varied and complex. Caregivers, whilst I am empathetic for how problematic regular vet visits can be, regarding time constraints, and emotionally and often financially, should be made aware of the importance of this. Many of the more popular breeds are commonly affected by skin issues. Long-term management demands time, education, and often financial investment. When these are lacking, welfare is compromised, and, as always, the life of the dogs we want alleviate from suffering.

Breeding, conformation and surgical birth

Potentially there is no bigger topic over the last 10 years in veterinary medicine than that of brachycephalic dogs (flat nosed) and breeding. The intersection of ethics, medicine, culture and financial reward make it the topic that all veterinary professionals know too well.

French Bulldogs, Pugs and English Bulldogs and the rest are among the UK's most popular breeds. They are also among the most consistently affected by conformation-linked disease.

BOAS (Brachycephalic Obstructive Airway Syndrome) – a disease that is directly resulting from congestion of the upper breathing apparatus due to genetics – spinal malformations, chronic eye infections and eyes issues are *extremely common*. Natural reproduction is deemed albeit problematic and somewhat inherently dangerous often. Planned caesareans are now frighteningly commonplace for these breeds. This is facilitated not only by veterinary surgeons but increasingly by commercial fertility clinics with questionable oversight. See *Britain's Puppy Boom: Counting the Cost* (BBC Three, 2021) as an introduction to the issues back during the pandemic.

I have seen and heard the many stories and realities of these issues. Female dogs are at every opportunity being bred and taken to vet clinics repeatedly to have caesareans, a difficult, labour-intensive and physiologically taxing intervention, even if you discredit the work that their bodies do to create and maintain life before and after this point. Many more illicit breeders do this with impunity whilst often failing to maintain good health or care. The bodies of these animals are secondary to the business side, which is aimed to produce as many puppies to sell as quickly as possible and to the highest bidder. Health considerations are paved over so as not to lower the cost and very little work is done to assess the homes to which they are going. This is not the case with all, but as we have seen, where there is profit to be had, there is an increasingly blurred line of what is acceptable and also how to enforce below standard care.

A breed selectively bred to require surgical birth is one whose welfare is fundamentally compromised. And here in the UK and in many Western countries such as the USA and in mainland Europe, we have allowed designer genetics, built around aesthetic only, to dominate the mainstream... and now we see the result.

Behavioural change as a red flag for pain

Pain is detail. Or more so, attention to it. I have seen caregivers pick up signs so faint and indistinct it is hard for me to not pass it off as superstition in some cases. However, I have learnt to never disregard the person who sees their dogs all day and night. The power in those changes can lead to some amazing discoveries. Behaviour is a powerful tool in my world.

Noise sensitivity, reactivity to handling and aggression during rest are all recognized behavioural manifestations of pain (Fagundes et al., 2018). The neurological literature supports this, describing behavioural outputs of pain ranging from startle responses to conditioned fear behaviours (Platt and Garosi, 2012).

It is here that collaboration with Clinical Animal Behaviourists becomes essential. This is why any good behaviourist will ask for proof that a full exam or at least some preliminary investigation has been done to rule out obvious signs of pain. As a vet, I can often locate pain or see changes in blood tests that may lead to diagnoses of ill-health. But it is often the behaviourist who does a more holistic investigation and sees the bigger picture of life for the dog: the dog who has stopped playing; the one who resists grooming; the one who barks when a child runs past. These are not isolated but part of a pattern of learnt behaviours. They are often stories much deeper than they seem.

Economic pressures and the veterinary dilemma

Pain management costs money and, to a greater extent, so does diagnostic imaging, blood testing, specialist referral and specific medicines. This reality can turn best-practice/gold standard medicine into a conversation about compromise. I believe that best practices often lead to a poor understanding of how results based on a happiness-first approach are undervalued or seen as less significant. We offer what is possible and that should be in discussion: open and two way. However, that conversation is in many cases difficult and doesn't always have the desired outcome... often due to frustration about lack of access to certain options.

Many caregivers are deeply committed to their animal's comfort but lack the means to pursue multimodal therapy: a situation I see regularly working with people experiencing homelessness. Others have the resources but struggle with understanding or motivation: a situation I see often when the cases are beyond the scope of buying a cure, or where cultural or philosophical differences rise to the surface. In both cases, communication is key. We must present pain not as an abstract diagnosis but as a lived reality, a lens through which the animal sees first, and then prioritize them as opposed to our preconceived notions of our preference. Happiness has to be the first consideration.

Reflections and the road ahead

We must see pain more actively, more honestly, more diligently, not simply as a symptom, but as a condition to remedy in itself. It is both the result and the disease that drives a lot of behaviour. That means it becomes a barrier to the wellbeing of our family members, even if they spread their fur all over our furniture.

All is not lost in vet world. Veterinary education is improving and growing. The pandemic and the now large influx of behavioural issues from lockdown puppies and relative increase in media attention to XL Bullies means we are talking more. Pain scoring, behavioural triage and owner-led observation tools are becoming mainstream and we are seeing a greater interest in behaviour as a topic. Collaboration between veterinary professionals, behaviourists and trainers is growing.

But we are not there yet.

We have to fight disinformation, opportunists on social media, who promote painful methods for compliance, alpha dog dialogue and intentionally avoid the vet because it is not cost effective. Suffering is a big umbrella that requires involvement from all sides and only with that commitment to improvement, will we see happier, healthier and stress-free dogs.

As professionals, we owe them that.

Pain and Wellbeing

As discussed in the above essay, when seeking to improve the wellbeing of your dog, or when assessing a client as a companion animal professional, pain and discomfort should always be considered. Conservative estimates have indicated that up to 80% of companion animals regularly experience some pain. Effective behaviour modification relies on a holistic approach to treatment and the failure to treat pain can cause treatment to delay or completely fail. The 2023 *Purina Institute Handbook of Canine and Feline Nutrition* (written by veterinarians) recommends a medical work-up not limited to a physical examination, urinalysis, fecal antigen testing and thyroid assessment for *all pets presenting with behavioural conditions* (Lenox *et al.*, 2023).

The presence of pain appears to lead to a less frequent experience of positive affect (Reaney *et al.*, 2017), meaning a reduction of positive mental state. Patients are also thought to experience a state of anxiety as commonly seen in humans with chronic pain (Camps *et al.*, 2012), oftentimes this is caused by the patient being fearful of re-injury (Dehghani *et al.*, 2004). Increased cognitive load due to discomfort might lead to reduced pain threshold and reduced ability to handle environmental stressors (Mills *et al.*, 2024).

Further factors leading to a compromised mental state include moderators not limited to a change in the sleep/wake cycle and an overall reduction of serotonin to the brain (Camps *et al.*, 2012).

While many problem behaviours are not caused by pain as a primary diagnosis, it is thought that a significant proportion of behavioural cases are made worse by the presence of pain (Mills *et al.*, 2020). Further research also considers paraesthesia ('abnormal prickling, tingling or burning sensation typically felt in the skin of the limbs') and dysesthesia ('abnormal painful sensations arising from innocuous cutaneous stimulation which are also often felt as a burning sensation that outlasts the stimulus') as moderators for behaviour problems (Mills *et al.*, 2024).

Chronic pain should be a consideration for a quality-of-life (QOL) assessment. There is no strict measure for QOL, and so each case should be tackled individually. Chronic pain requires the animal to appraise the world around them, and this creates a worldview that in some cases can be overwhelmingly negative.

> It is not an anatomical, physiological or pharmacological state; rather, it is a matter of how internal and external conditions are perceived by that individual animal, or how it 'feels'.
>
> (Davis *et al.*, 2019)

To add to this, the animal is likely not just to be living in pain (and coping with the emotional stressors that accompany it), but they may also be coming into conflict with humans and other animals due to pain-related behaviour changes. In many cases, individuals are subjected to aversive or painful training methods (as described in earlier chapters), which will only further exacerbate their pain.

Pain and Behaviour

There are certain scenarios in which caregivers are more likely to recognize behavioural signs of pain in their dogs. Dogs that exhibit postural or movement-related changes are typically investigated for pain (Ahu *et al.*, 2023) as well as those that show reduced activity levels. An example given in a study by Davis *et al.* (2019) is as follows:

> We would come home, she would be really excited and stand up on her hind legs and put a paw on each side of your hips and literally give you a hug. Over time, she lost that range of motion and she doesn't stand up anymore.

However, less obviously linked behaviour changes can indicate pain. Behaviours linked to painful conditions include defensive or reactive behaviours, increased clinginess, abnormal repetitive behaviours and refusal of walks to name a few (Mills *et al.*, 2020). These behaviours may precede behaviours typically associated with pain such as limping and yelping (Canine Arthritis Management, 2025), and in fact Canine Arthritis Management (CAM) advises that by the time a dog is yelping they are in a significant amount of pain and discomfort.

When asking a caregiver or veterinary professional to consider pain in a behaviour case, there can be some obstacles. For example, it is sometimes assumed that young dogs are unlikely to be experiencing chronic pain, or conversely that older dogs will inevitably experience some pain as a symptom of ageing (Mills *et al.*, 2024). CAM states that around 35% of dogs in all age groups will experience chronic pain, and 80% of dogs over 8 years of age. Further still, caregivers will often find that their dogs will continue to partake in activities that they find pleasurable (for example, chasing a ball), struggling to conceive how a dog in chronic pain could choose to do such things (Mills *et al.*, 2024). It is thought that a good mood or overall optimistic personality can be enough for a dog to mask their pain (Reaney *et al.*, 2017). Having reservations should not prevent a caregiver or practitioner from investigating pain when faced with any behavioural changes.

Identifying pain in a behaviour case

While it should go without saying that a qualified veterinarian is the professional tasked with identifying and treating pain in a behaviour case, there are some signs that caregivers and behaviour practitioners can look for to raise suspicions. As well as this, all parties should consider the variety of conditions that cause pain in animals. Not only are musculoskeletal conditions a consideration, but pain can be caused by chronic gut conditions, skin issues and allergies to name some examples. Examples of questions a behaviourist might ask are included in Appendix A .

Behaviour and musculoskeletal pain

Fast or sudden onset of aggressive behaviour can be an indicator of pain. This is particularly pertinent when the aggression appears impulsive, when the dog is aggressive as a result of physical manipulation, or when they become defensive in their posture (Camps *et al.*, 2012). These behaviours are intended to 'avoid physical contact that may cause further injury' (Rutherford, 2002). For example, a dog that previously tolerated having their feet cleaned may quickly become aggressive and have a very low tolerance before snapping or biting. In a case such as this the dog is not only showing impulsive aggression but is reacting to their body being physically manipulated.

Fear responses, particularly related to noise, have been linked to pain in dogs (Fagundes *et al.*, 2018). The 2018 paper referred to studies in humans that showed links between painful conditions and avoidance behaviours, and so it is recommended that dogs with noise phobias are investigated for pain. Though musculoskeletal pain is thought to be a primary concern, questions about gut and thyroid issues are also raised.

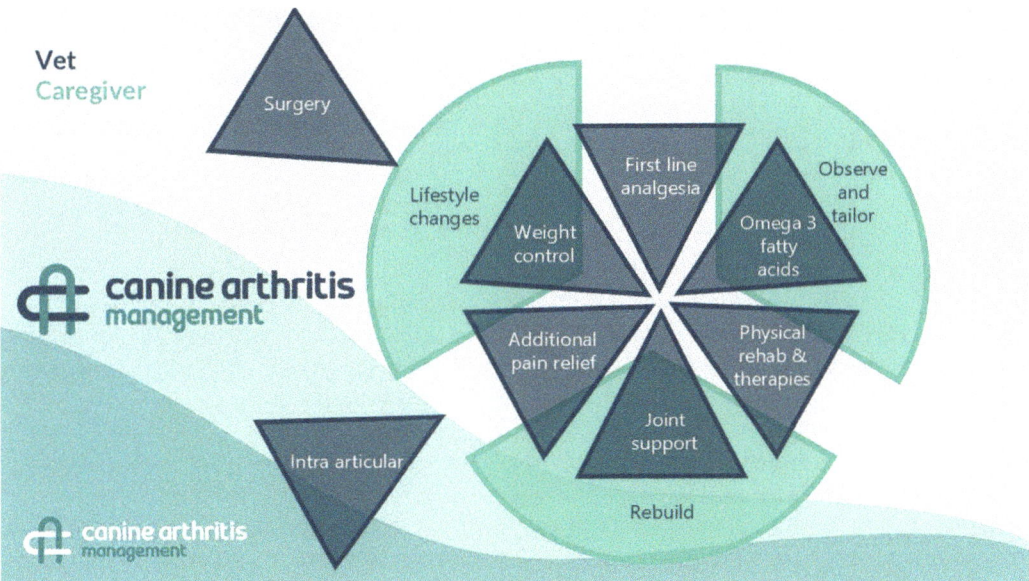

Fig. 9.1. CAM Multimodal Management model (PDSA, 2024; Canine Arthritis Management, 2025) Used with permission.

The Link Between Physical and Emotional Wellbeing

As a Clinical Animal Behaviourist, I have personally treated dogs for a presenting complaint of aggression and upon veterinary examination they have been found to have hip and elbow dysplasia, slipped vertebrae, neck injuries and other physical conditions. A 2012 study (Camps *et al.*, 2012) surveyed 12 aggressive dogs suffering from painful conditions and found that 75% (9 of 12) were suffering from musculoskeletal conditions (hip dysplasia and elbow osteoarthritis).

As stated earlier, dogs with suspected pain should always be seen by a qualified veterinary professional. Caregivers and behaviour professionals can also access First Aid and management advice via the CAM website. With these interventions and using the 'CAM Multimodal Management model' (Fig. 9.1), painful conditions can be managed.

Case vignette: Mara

Mara was referred to me for a sudden onset of aggressive behaviour towards other dogs. At the time of consultation Mara was 22 months old. Mara is a neutered female Rottweiler.

Prior to seeing me, Mara had attended a reputable puppy class and had regular social visits to a small daycare centre. There was no previous history of aggression towards other dogs or towards people. Generally speaking, Mara had always been a well-mannered dog. At the time of Mara's referral she had started grumbling towards dogs that she'd known for a long time as well as unknown dogs she would encounter on walks.

Mara had a history of food intolerances that had led to some skin conditions and recurring ear infections. Mara had also become increasingly worried during her vet visits.

Fig. 9.2. Mara's behaviour changed as a result of chronic ear infections and musculoskeletal issues. Image is author's own.

Through consultation, it was determined that there was no obvious cause of Mara's recent change in behaviour. Mara was prescribed a desensitization and counter-conditioning programme to be completed from a distance around dogs as well as some exercises to engage with her handler rather than fixating on incoming dogs. Further to this, Mara visited the vet for a review of her physical health.

Mara's vet decided to sedate her and flush her ears. Both ears were found to be severely infected and required treatment.

With Mara's medical issues under control as well as the behaviour modification protocols underway, Mara slowly went back to parallel walking with other dogs before building up to free play in some appropriately chosen settings. Her painful condition has caused some memory of pain and so Mara's caregivers often still opt for her to be muzzled during initial greetings as a precaution.

More recent updates from Mara's family revealed an increase in 'grumpiness' around other dogs when in a play context, particularly those who are bouncy or invasive of personal space. Going back once again to the vet revealed musculoskeletal abnormalities including a condition requiring shoulder surgery. Since recovering from surgery Mara has once again been able to go back to her friends (Fig. 9.2).

Behaviour and gastrointestinal issues

A common cause of behaviour change in dogs is due to changes in the gastrointestinal system, otherwise known as the 'gut'. These changes can be caused by something as simple as an allergy or an intolerance, and can be accompanied by other strange behaviours such as eating non-food items like grass and stones (Radosta, 2023).

The 'gut microbiome' is the collective name for the various microorganisms in the gastrointestinal system. Changes in the microbiome have been found in relation to various diseases and disorders, not least those in the neurological system (Suchodolski, 2022) and practitioners often refer to the 'gut–brain axis', 'a bi-directional communication of the nervous, endocrine, and immune systems' (Cannas *et al.*, 2021). Studies have shown that the gut microbiome somewhat influences aggression (Kirchoff *et al.*, 2019), memory (Kubinyi *et al.*, 2020) and emotions related to stress and anxiety (Cannas *et al.*, 2021).

Luckily, adjusting a dog's diet has been shown to complement some behaviour modification treatment plans (Radosta, 2023). Though these adjustments are unlikely to completely cure behavioural disorders, it has been suggested that they reduce the amount of stress a dog is exposed to. Veterinarians and canine nutritionists can help to formulate diet changes in order to support behaviour modification.

Behaviour and cognitive health

While not necessarily a painful condition, there are both behaviour and welfare implications for declining cognitive health in dogs. The medical condition known as 'Cognitive Dysfunction Syndrome' (CDS) is suspected to be present in 30% of dogs aged 11–12 years and 60% of dogs aged 15+years (Volk, 2023). This disease is known to closely resemble Alzheimer's Disease in humans (Fast *et al.*, 2013).

Studies have established key behavioural indicators of Canine Cognitive Dysfunction Syndrome (CCDS) including changes to the sleep/wake cycle, changes to social interaction, signs of disorientation and generalized anxiety (Fast *et al.*, 2013).

When left untreated, CCDS can be considered a huge compromise to canine emotional wellbeing due to the noticeable levels of anxiety in these patients. It is thought possible to improve the cognitive functioning of patients with CCDS through diet (Volk, 2023) and the use of supplements (Heath *et al.*, 2007) and so veterinary referral to a canine nutritionist is one option. As well as this, many of the strategies applied in Chapter 5 regarding 'trauma-informed care' can be beneficial in improving a dog's emotional wellbeing as safety and predictability are key in both contexts.

Improving Welfare During Handling

Handling in the veterinary clinic

A high percentage of pet dogs experience apprehension or stress when visiting the veterinary clinic. Surveys have shown that around 60% of dogs are hesitant or unwilling to enter the clinic (Carroll *et al.*, 2022), with 13.3% of patients having to be physically dragged into the clinic (Lloyd, 2017), and that approximately 78% of patients exhibit fearful behaviours when put onto the examination table (Döring *et al.*, 2009). Patients are more likely to display such fearful behaviour at the vet if they have been trained using positive punishment (Rosengren *et al.*, 2023). Not only is this fearful behaviour at the veterinary clinic a welfare concern due to the stress experienced by the dog, but it presents limitations for the veterinarian performing the exam (Stellato *et al.*, 2019; Calder, 2024).

One option for mitigating this problem is to use an online veterinary care service, many of which are becoming available in the UK. These services are able to advise on around 70% of medical complaints (Vetsy, 2025) and are a great option for high-welfare medical care. However, not all veterinary care can be delivered remotely and so it falls to those in the veterinary clinic to support animals and reduce the levels of stress that they experience. As little as one negative experience at the veterinary clinic can impact a dog's view of future visits (Lloyd, 2017).

First, caregivers and professionals can apply 'Low Stress Handling®' techniques.

Low Stress Handling®

Low Stress Handling Restraint and Behavior Modification of Dogs & Cats: Techniques for Developing Patients Who Love Their Visits was a groundbreaking book written by Dr Sophia Yin in 2009. This book shared core principles for reduction of stress during veterinary clinic visits and, when applied, said techniques have been largely successful (Scalia *et al.*, 2017). Starting with minimal restraint, stricter measures are only applied based on analysis of the dog's behaviour and any risk to the professional performing the procedure (Carroll *et al.*, 2022).

A lot of low stress methods seek to give animals an element of control and predictability, since the absence of these things is one reason animals are likely to feel stress (Lloyd, 2017).

Dr Yin shares 10 key principles (taken verbatim from her 2009 book):

1. Start with a comfortable environment.
2. Keep the animal from pacing, moving nervously or excitedly, squirming, or suddenly trying to escape.
3. Support the animal well by having your hands, arms and body positioned appropriately. The pet should not feel as if he will fall or is off balance.
4. Be aware that physically positioning animals or asking them to perform behaviors when they are nervous, scared or confused can cause them to resist handling. Thus, when they act confused or scared, it's important to move deliberately and slowly to avoid a reflex resistance or escape behavior.
5. Know how to place your hands and body to control movement in any direction.
6. Wait until the pet is relaxed before starting a procedure.

7. Use the minimum restraint needed for the individual.
8. Avoid prolonged (more than 2 s) or repeated fighting or struggling.
9. Use distractions and rewards when appropriate.
10. Adjust your handling based on the animal and his response to restraint and learn to adjust your sample-taking technique.

While the above guidelines are mostly applicable to veterinary professionals, caregivers and other pet professionals can adopt some principles in order to make handling more comfortable.

Optimizing the environment

Some studies have shown that adjusting waiting room protocols can positively impact the patient. For example, it is recommended to bring a familiar bed or blanket (Calder, 2024) and allow the animal time to acclimatize in the waiting room (Hernander, 2008). Since the process of weighing the animal is considered highly stressful, it is recommended to do this right at the beginning and allow time to recover before consultation (Hernander, 2008). Additionally, this way the animal does not become classically conditioned to weighing as a predictor of going into the consultation room.

Supporting dogs through consultation

Try to locate a vet who uses Low Stress Handling® techniques. Fig. 9.3 demonstrates the Low Stress Handling® algorithm recommended for veterinarians. Not only is there evidence that these techniques are beneficial to patients, studies have shown that caregivers prefer for their dogs to be examined in a minimally invasive way (Carroll *et al.*, 2022). Further materials are available via *Cattle Dog Publishing* online.

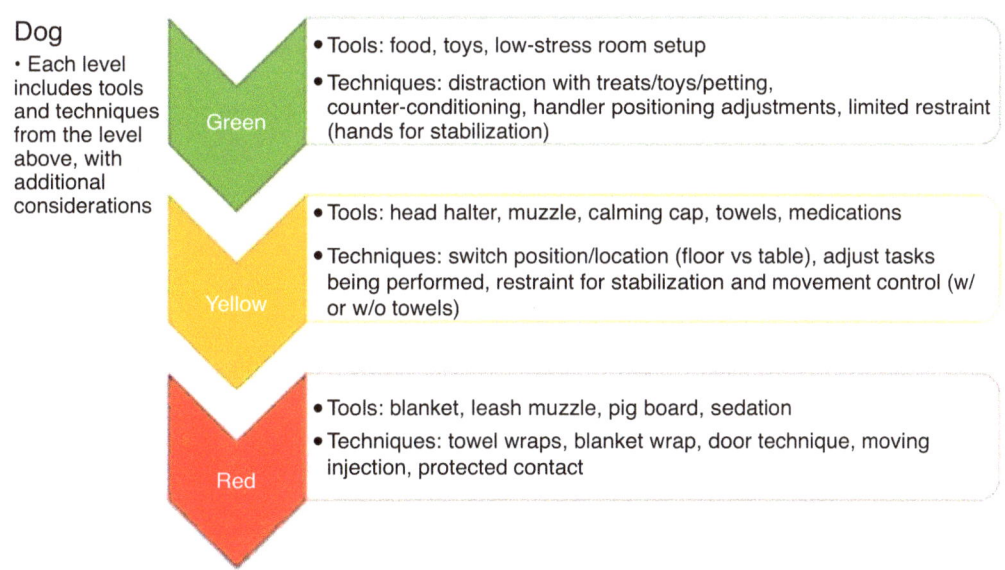

Fig. 9.3. Low Stress Handling® algorithm (Calder, 2024). Shared with permission.

Once in the consultation room, many dogs will be responsive to food rewards (Lloyd, 2017). It is also preferable for dogs to be examined on the floor or in the lap of their owner since being placed on the table is a known cause for concern. In cases of extreme fear responses to the clinic, caregivers are encouraged to request their consultation outside. Much like the waiting area, animals should be given time to explore their environment before physical examination occurs (Rosengren et al., 2023).

Procedures necessary to the examination will be determined by the consulting veterinary professional, but dogs can be desensitized to handling at home in order to make these interactions easier. An example of how to do so is by teaching the dog a co-operative care protocol. If contacted beforehand some vets will be open to working through a co-operative care protocol in the clinic which reduces stress significantly for the patient. Studies have shown that 'physiological manipulation' is the greatest source of stress for dogs in the clinic, and so particular attention should be directed here (Rosengren et al., 2023).

Handling at the grooming salon

Another experience many dogs find highly aversive is going to the dog groomer. The 2024 National Dog Survey by Dogs Trust (UK) collected data on 430,000 dogs, finding that more than 40% of these dogs are crossbreeds. Of course, many purebred dogs also require coat maintenance, and it is unclear how many of the 40% are crossbreeds that require coat care. However, with an obvious rise in poodle crossbreed popularity (contributing in no small way to the 40%) there is a mounting pressure on the dog-grooming sector.

Appropriate grooming is paramount to physical wellbeing. Dogs can experience a variety of health consequences in the absence of suitable grooming including matting (in some circumstances leading to skin irritation, constricted blood flow and eventually soft tissue death), postural and gait abnormalities due to poor nail health (McDonald et al., 2022) and finally, grooming appointments are a great way to recognize other medical conditions early.

Traditionally, dog grooming has required a high level of restraint, something we know is highly stressful for many dogs. Not only that, but the small number of studies that have measured stress in the dog grooming setting have found the grooming salon itself to be a stressor (Mariti and Bein, 2015). As well as anticipating the grooming procedure, dogs are likely to be stressed by being handled by strangers, having to deal with other dogs and being exposed to a variety of novel equipment and supplies (Ferreira et al., 2022).

Choosing a high-welfare salon

There are some characteristics that would make one grooming salon superior to another in terms of welfare. Dogs will benefit from the following:

- Few or no other dogs being groomed at the same time and in the same space.
- Reduced time spent restrained or caged.
- Longer appointments allowing patient handling by groomers.
- Quieter or noise-suppressed equipment.

As well as this, salon modifications designed for comfort would likely make a large difference to the dog in question as this Brazilian study revealed:

Successful escape attempts [by animals being groomed] resulted in falls from the bathtub or drying table. In some cases, the animals were temporarily hanged by the leash attached to the neck collar until the employees were able to put the dogs back to the handling areas. This could have caused severe injuries and demonstrates the discomfort of animals with the handling to which they were subjected.

(Ferreira *et al.*, 2022)

For example, the difference between using a non-slip mat and not using one could be massive for any dog needing a trip to the groomer. Simply reducing the amount of slippery surface in the bath could prevent escape attempt injuries such as the ones described above. Dogs in these situations could be supported further by desensitization at home.

Guidance for groomers

From a welfare and behaviour perspective, there are several things that groomers can implement in order to benefit their customers. However, in order for a dog's experience to be as positive as possible their caregivers have the responsibility of desensitizing their pet to handling as well as staying on top of any potential health issues.

Groomers can implement the above guidance (quieter environment, fewer dogs, less time in restraint or cages) in their salon as ways to decrease the overall stressors that their dogs are exposed to. Investing in higher quality dryers as well as non-slip mats can also prevent injuries related to panic.

When working with dogs in the salon, groomers will probably benefit from:

- The use of food rewards as a way to build positive associations.
- Using luring (asking dogs to follow a food reward) to ask dogs to move rather than forcibly handling them.
- Providing breaks in between handling.
- Working on the floor where possible.
- Behaving calmly and patiently with slow movements rather than rough and jerky movements.
- Using as few restraints as possible whilst still keeping the dog and themselves safe.

It is also worth remembering that the greater a groomer's understanding of canine communication and body language, the easier it will be to see defensive or aggressive behaviour starting. So, knowledge in this area is always recommended.

Desensitization and Co-operative Care

In the majority of conversations I have had with vets and dog groomers, there is an agreement that more can be done at home to prepare dogs for any sort of formal handling. The kindest and most effective way is through systematic desensitization, as mentioned frequently throughout this book.

Applying this technique to veterinary visits has been observed as a successful fear reduction strategy in small studies (Stellato *et al.*, 2019). Therefore, caregivers should still be advised to proceed with systematic desensitization for formal handling. One high-welfare method of doing so is through a practice known as 'co-operative care'.

What is co-operative care?

> Cooperative care is a means of interacting with your dog in such a way that they become a cooperative partner in the process. As well as using positive reinforcement to create an enjoyable grooming, veterinary, or husbandry experience, you are also giving them a voice in the process by teaching them how to opt-in and to opt-out.
>
> (Pachel, 2023)

Knowing that handling is often stressful for dogs, many practitioners choose to teach their dog a consent-based method known as 'co-operative care'. Providing the animal with an element of control in a situation is an excellent way to reduce fear and increase tolerance (Dixon *et al.*, 2018). Dixon goes on to share the benefits of the practice, 'many people assume that giving an animal a sense of control would just result in the animal refusing to participate in the unpleasant experience—the opposite is true when cooperative care is properly trained'.

Limitations

Whilst this should be the 'gold standard' of care for dogs (particularly those significantly distressed by handling), there are limitations concerned with consent-based or low-stress handling. For example, Stellato *et al.* (2019) tried to implement a desensitization protocol for dogs experiencing fear of the veterinary clinic but unfortunately found that owner compliance was very poor (also seen in a study by Wess *et al.*, 2022). Further to this, if the dog is particularly anxious, is in pain or experiencing cognitive decline there may be limited capacity for behaviour modification in this case (Dobson, 2012). Finally, some procedures are considered so aversive that co-operative care will not be an option for that dog. This was demonstrated in the Wess *et al.* (2022) study, which saw no benefit of using co-operative care with dogs undergoing a rectal temperature check.

Teaching a co-operative care protocol

Despite the timely nature of teaching a co-operative care protocol as well as the expressed limitations, it remains a high welfare way to perform husbandry and handling with animals. As such, it is advisable to teach co-operative care to any animal reluctant or resistant to handling. It can also be used as a preventative measure on animals who have not had any negative experiences.

Follow the below steps to teach a co-operative care protocol to your own dog or to a client you are working with. Since handling is a stressful experience for many dogs, the protocol may not always work perfectly. Continue to be consistent and patient for meaningful results. If handling is required and you cannot fully comply with this protocol (for example, treatment is time sensitive) do not try to do so through co-operative care as you cannot risk rushing and 'poisoning' your protocol. As written in Erin Jones' 2024 book *Constructing Canine Consent*: 'only ask if you can accept **no** for an answer'.

Establish the behaviour you would like your dog to 'consent' to

Decide which behaviour you would like your dog to perform in order to perform your handling exercises. Well known examples include 'The Bucket Game' designed by London trainer and behaviourist Chirag Patel, and a simple chin rest. Whatever behaviour you choose, the dog should perform said behaviour as a way to 'opt in' to training, and cease to perform the behaviour as a way to 'opt out'.

This example will follow the use of a chin rest behaviour (Fig. 9.4). To clarify, the dog should perform the chin rest when they are happy to proceed with handling and lift their head up (stop the behaviour) when they want handling to stop. It's very important to establish this reliable non-aggressive way of asking the handler to stop and to teach the animal that they have the option to say 'no' (Dixon *et al.*, 2018).

Shape the behaviour and reward for it

A chin rest can be performed whilst the dog is in a seated, standing or laying down position. For ease, choose one position and stick to it. This example will show a chin rest from a laying down position.

Use something that the dog can recognize as a target. A good example is a rolled-up towel or cushion. With a treat in your hand, lure the dog's muzzle down onto the target. Once the muzzle is in the right place mark the behaviour (for example with a clicker) and reward. Continue to repeat this behaviour.

It is important to note that the dog may lift their head very briefly to take the reward but should put their chin straight back down so as to continue consenting.

Build duration

Once the desired behaviour has been learned the criteria should increase to a chin rest with duration. Aim for a couple of extra seconds at a time.

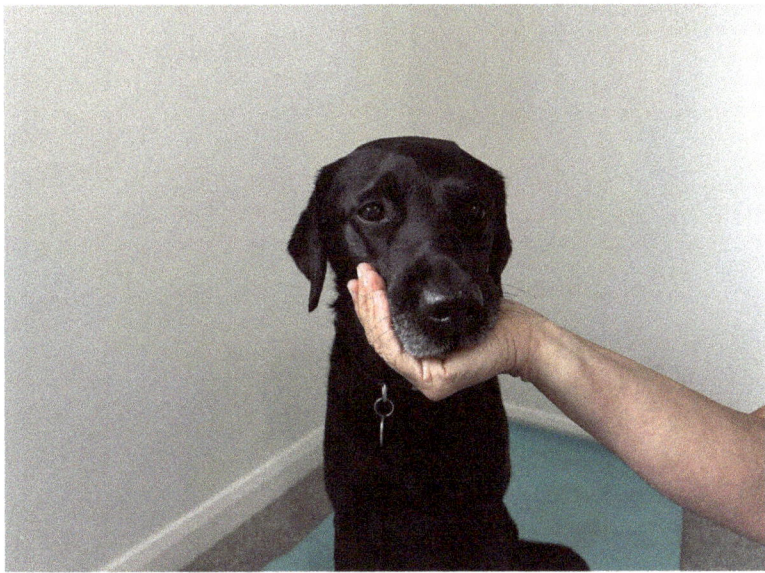

Fig. 9.4. Coconut learns to offer a chin rest with duration. Image supplied by Tamsin and Coconut Durston.

The Link Between Physical and Emotional Wellbeing

In order to build duration in the chosen behaviour:

- Delay the reward slightly.
- Reward quickly and consistently, keeping the rewards coming as the behaviour is held in place.
- You can add a pattern game if you like to count aloud while the position is held.

Teach a release cue

With the building of duration comes the establishment of a clear release cue. This cue tells the dog that the exercise is completed and that they can break the behaviour. Try something like 'free' and throwing a treat away from the dog.

You can use this to end the session or simply to offer the dog a break.

Part of teaching a release is also to cease any training when the dog lifts their head away from the target. This enables them to understand that they are saying 'no' without the need for any aggression.

Start to introduce handling practice

Once the dog has reliably learned the behaviour, is offering it with duration and is able to stay in position until released, handling exercises can start to be introduced.

Taking nail clipping as an example: introduce the exercise over a *number of sessions* while maintaining the consent-based protocol (Fig. 9.5). A gradual process of desensitization may look something like this:

- Place the nail clippers somewhere visible to the dog but do not use them.
- Gently manipulate the dog's nails using your hand (e.g. wiggle each nail).
- Using the nail clippers gently tap the dog's nail.
- Cut one nail (proceed based on rate of success!).

Suggested desensitization exercises

When a dog is at the stage they are ready to learn some handling exercises, it is worth considering what the priorities are. In the 2018 book *Cooperative Care: Seven Steps to Stress-Free Husbandry*, Dr Deborah Jones (2018) shares '10 essential' exercises for effective co-operative care:

1. Chin rest
2. Lie on side
3. Restraint
4. Muzzle training
5. Foot handling
6. Mouth handling
7. Taking medication
8. Injection/Blood draw
9. Eye exam
10. Ear exam

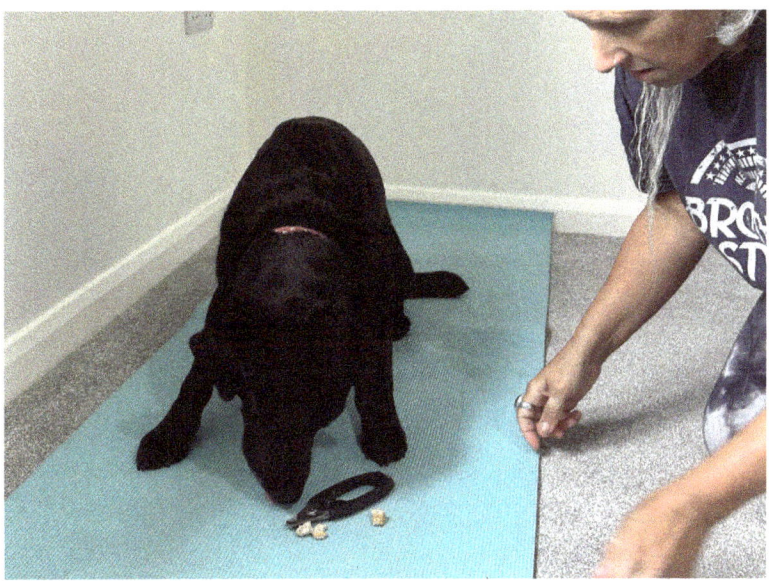

Fig. 9.5. Sometimes dogs need to be allowed to explore 'scary' equipment in their own time before we can physically use it with them. Here Coconut explores some nail clippers in order to associate them with rewards. Image supplied by Tamsin and Coconut Durston.

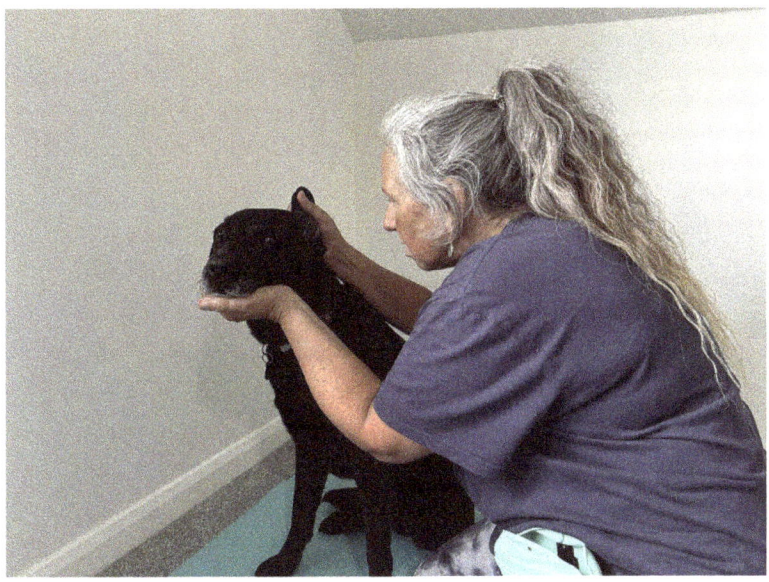

Fig. 9.6. Coconut practises an ear check and clean. She has the option to lift her chin if she wishes to stop the exercise. Image supplied by Tamsin and Coconut Durston.

All exercises should be introduced gradually as approximations before trying to perform them in full (Figs 9.6 and 9.7). If a particular exercise proves problematic pain should be investigated as a primary cause.

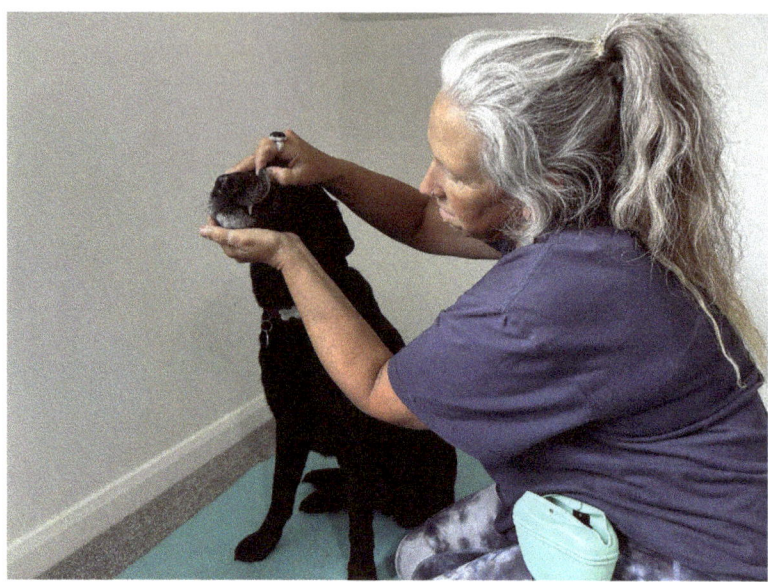

Fig. 9.7. Coconut practises a consensual teeth check. Image supplied by Tamsin and Coconut Durston.

Conclusion

The idea of veterinary handling or grooming strikes fear in the heart of many caregivers and professionals alike. Knowing how stressful many dogs find these visits puts the individual in a difficult moral position: knowing that they have to make sure their dog is healthy but also understanding that it is a highly aversive experience for many. Furthermore, the added pressure of behavioural problems potentially arising from pain or discomfort makes the urgency of veterinary visits more pressing, and failing to investigate these early signals can lead to significant breakdown in relationships.

Luckily, research has shown some clear ways to make these visits less stressful for dogs. In particular, adopting a consent-based approach and teaching the dog a co-operative care protocol has shown promise in many studies. While the process of teaching such a process is repetitive and often long-winded, it can be the key to lifting the pressure off of all parties involved and getting that dog the help that they need.

Disclaimer: Please be aware that this chapter (aside from Fabian's essay) is written by a non-vet. If you have any doubts about the health of your dog you should contact a qualified veterinarian. Further to this, your chosen trainer or behaviourist should take all veterinary guidance on board in order to treat behaviour problems thoroughly and holistically.

References

Ahu, D., Durmus, A., Begum, S., Sevim, I., Hakan, O. *et al.* (2023) Dog owners' recognition of pain-related behavioral changes in their dogs. *Journal of Veterinary Behavior* 62, 39–46.

BBC Three (2021) *Britain's Puppy Boom: Counting the Cost.* Available at: https://www.bbc.co.uk/programmes/p09nl348 (accessed 24 July 2025).

Calder, C.D. (2024) Redefining low stress handling®. Available at: https://www.cattledogpublishing.com/blog/whats-new-in-low-stress-handling (accessed 15 July 2025).

Camps, T., Amat, M., Mariotti, V.M., Brech, S. and Manteca, X. (2012) Pain-related aggression in dogs: 12 clinical cases. *Journal of Veterinary Behavior* 7(2), 99–102.

Canine Arthritis Management (2025) Arthritis in Dogs - The Basics. Available at: https://caninearthritis.co.uk/what-is-arthritis/arthritis-in-dogs-the-basics/ (accessed 15 July 2025).

Cannas, S., Tonini, B., Belà, B., Prinzio, R., Pignataro, G. *et al.* (2021) Effect of a novel nutraceutical supplement (Relaxigen Pet Dog) on the fecal microbiome and stress-related behaviors in dogs: A pilot study. *Journal of Veterinary Behavior* 42, 37–47.

Carroll, A.D., Cisneros, A., Porter, H., Moody, C. and Stellato, A.C. (2022) Dog owner perceptions of veterinary handling techniques. *Animals* 12, 11. DOI: 10.3390/ani12111387.

Coatesworth, J. (2019) *Small Animal Dermatology: What's Your Diagnosis?* John Wiley & Sons, Hoboken, New Jersey.

Davis, K.N., Hellyer, P.W., Carr, E.C.J., Wallace, J.E. and Kogan, L.R. (2019) Qualitative study of owner perceptions of chronic pain in their dogs. *Journal of the American Veterinary Medical Association* 254, 88–92.

Dehghani, M., Sharpe, L. and Nicholas, M.K. (2004) Modification of attentional biases in chronic pain patients: A preliminary study. *European Journal of Pain* 8(6), 585–594.

Dixon, S., Fraser, L. and Edlund, S. (2018) What is cooperative care? *IAABC Foundation Journal*. Available at: https://journal.iaabcfoundation.org/cooperative-care/ (accessed 15 July 2025).

Dobson, J. (2012) Vet Times Fears, phobias and anxiety disorders in cats and dogs. Available at: https://www.vettimes.co.uk (accessed 15 July 2025).

Dogs Trust (2024) Welcome to the Results of the National Dog Survey 2024. Available at: https://www.dogstrust.org.uk/downloads/Dogs_Trust_NDS_Report_2024__.pdf (accessed 10 July 2025).

Döring, D., Roscher, A., Scheipl, F., Küchenhoff, H. and Erhard, M.H. (2009) Fear-related behaviour of dogs in veterinary practice. *The Veterinary Journal* 182(1), 38–43.

Fagundes, A.L.L., Hewison, L., McPeake, K.J., Zulch, H. and Mills, D.S. (2018) Noise sensitivities in dogs: An exploration of signs in dogs with and without musculoskeletal pain using qualitative content analysis. *Frontiers in Veterinary Science* 5, FEB. DOI: 10.3389/fvets.2018.00017.

Fast, R., Schütt, T., Toft, N., Møller, A. and Berendt, M. (2013) An observational study with long term follow up of canine cognitive dysfunction: Clinical characteristics, survival, and risk factors. *Journal of Veterinary Internal Medicine* 27(4), 822–829.

Ferreira, M., Rodriguez, M.A.P., Santos Oliveira, L.L., Albuquerque Maranhão, C.M., Oliveira, N.J.F. *et al.* (2022) Stress in dogs during grooming in a pet shop. *Revista Brasileira de Zootecnia* 51, e20200154. DOI: 10.37496/RBZ5120200154.

Hall, E.J., Williams, D.A. and Kathrani, A. (eds) (2020) *BSAVA Manual of Canine and Feline Gastroenterology*, 3rd edn. British Small Animal Veterinary Association, Gloucester, UK.

Heath, S.E., Barabas, S. and Craze, P.G. (2007) Nutritional supplementation in cases of canine cognitive dysfunction—A clinical trial. *Applied Animal Behaviour Science* 105(4), 284–296.

Hernander, L. (2008) Factors influencing dogs' stress level in the waiting room at a veterinary clinic. Sweden: Swedish University of Agricultural Sciences.

IASP (International Association for the Study of Pain) (2024) Terminology. Available at: https://www.iasp-pain.org/resources/terminology/?utm_source=chatgpt.com (accessed 15 July 2025).

Jones, D.A. (2018) *Cooperative Care: Seven Steps to Stress-Free Husbandry*. Dogwise, Wenatchee, Washington.

Jones, E. (2024) *Constructing Canine Consent*. CRC Press, Boca Raton, Florida.

Kirchoff, N.S., Udell, M.A.R. and Sharpton, T.J. (2019) The gut microbiome correlates with conspecific aggression in a small population of rescued dogs (*Canis familiaris*). *PeerJ* 2019(1), 6103. DOI: 10.7717/peerj.6103.

Kubinyi, E., Bel Rhali, S., Sándor, S., Szabó, A. and Felföldi, T. (2020) Gut microbiome composition is associated with age and memory performance in pet dogs. *Animals* 10(9), 1–10. DOI: 10.3390/ani10091488.

Lenox, C., Corbee, R.J. and Sparkes, A. (eds) (2023) *Purina Institute Handbook of Canine and Feline Clinical Nutrition*, 2nd edn. Embark Consulting Group, LLC, Medfield, Massachusetts.

Lloyd, J.K.F. (2017) Minimising stress for patients in the veterinary hospital: Why it is important and what can be done about it. *Veterinary Sciences* 4(2), 22. DOI: 10.3390/vetsci4020022.

Mariti, C. and Bein, S. (2015) Evaluation of dog welfare before and after a professional grooming session. *Dog Behaviour* 1, 1. DOI: 10.4454/db.v1i1.2.

McDonald, S.E., Doherty, C., Sweeney, J., Kisiel, L., Matijczak, A. *et al.* (2022) Barriers to and facilitators of pet grooming among clients served by a subsidized grooming service program. *Frontiers in Veterinary Science* 9, 1021707.

Mills, D.S., Demontigny-Bédard, I., Gruen, M., Klinck, M.P., McPeake, K.J. *et al.* (2020) Pain and problem behavior in cats and dogs. *Animals* 10(2), 318. DOI: 10.3390/ani10020318.

Mills, D.S., Coutts, F.M. and McPeake, K.J. (2024) Behavior problems associated with pain and paresthesia. *Veterinary Clinics of North America - Small Animal Practice* 54, 55–69. DOI: 10.1016/j.cvsm.2023.08.007.

Monteiro, B.P., Lascelles, B.D.X., Murrell, J., Robertson, S., Steagall, P.V.M. *et al.* (2023) 2022 WSAVA guidelines for the recognition, assessment and treatment of pain. *Journal of Small Animal Practice* 64(4), 177–254. DOI: 10.1111/jsap.13566.

My Vetsy, M. (2025) About us. Available at: https://www.myvetsy.co.uk/about-us (accessed 12 June 2025).

Pachel, C. (2023) Adaptil. *Cooperative Care for Dogs: A Happy Dog Expert Explains*. Available at: https://www.adaptil.co.uk/blogs/news/cooperative-care-for-dogs-a-happy-dog-expert-explains/ (accessed 15 July 2025).

PDSA (2024) PDSA Animal Wellbeing (PAW) Report. Available at: https://www.pdsa.org.uk/what-we-do/pdsa-animal-wellbeing-report/paw-report-2024 (accessed 10 July 2025).

Platt, S. and Garosi, L. (eds) (2012) *Small Animal Neurological Emergencies*. Manson Publishing, London.

Platt, S.R. and Olby, N.J. (eds) (2013) *BSAVA Manual of Canine and Feline Neurology*, 4th edn. British Small Animal Veterinary Association, Gloucester, UK.

Radosta, L. (2023) Behavioral disorders in dogs and cats. In: Lenox, C., Corbee, R.J. and Sparkes, A. (eds) *Purina Institute Handbook of Canine and Feline Clinical Nutrition*, 2nd edn. Embark Consulting Group, LLC, Medfield, Massachusetts.

Reaney, S.J., Zulch, H., Mills, D., Gardner, S. and Collins, L. (2017) Emotional affect and the occurrence of owner reported health problems in the domestic dog. *Applied Animal Behaviour Science* 196, 76–83.

Rosengren, S.S., Viktor, J. and Hänninen, L. (2023) *Stress Free Handling Methods of Dogs in the Veterinary Practice*. University of Medicine, Budapest.

Rutherford, K.M.D. (2002) Assessing pain in animals. *Animal Welfare* 11(1), 31–53.

Scalia, B., Alberghina, D. and Panzera, M. (2017) Influence of low stress handling during clinical visit on physiological and behavioural indicators in adult dogs: A preliminary study. *Pet Behaviour Science* 4, 20–22.

Stellato, A., Jajou, S., Dewey, C.E., Widowski, T.M. and Niel, L. (2019) Effect of a standardized four-week desensitization and counter-conditioning training program on pre-existing veterinary fear in companion dogs. *Animals* 9(10), 767. DOI: 10.3390/ani9100767.

Suchodolski, J.S. (2022) Analysis of the gut microbiome in dogs and cats. *Veterinary Clinical Pathology* 50(S1), 6–17. DOI: 10.1111/vcp.13031.

Volk, H. (2023) Brain disorders in dogs and cats. In: Lenox, C., Corbee, R.J. and Sparkes, A. (eds) *Purina Institute Handbook of Canine and Feline Clinical Nutrition*, 2nd edn. Embark Consulting Group, LLC, Medfield, Massachusetts.

Wess, L., Böhm, A., Schützinger, M., Riemer, S., Yee, J.R. *et al.* (2022) Effect of cooperative care training on physiological parameters and compliance in dogs undergoing a veterinary examination – A pilot study. *Applied Animal Behaviour Science* 250. DOI: 10.1016/j.applanim.2022.105615.

Yin, S. (2009) *Low stress handling, restraint and behavior modification of dogs and cats: Techniques for patients who love their visits*. CattleDog Publishing, Davis, California.

Appendix A

Behaviour questions

The following questions can be used to establish whether any behaviour change might be linked to pain. They can be asked within consultation or beforehand in a history form, and then the information gathered should be passed on to the referring veterinarian. These questions are by no means exhaustive but are intended to be a starting point.

Diet

1. Is your dog motivated by food?
2. What is their feeding routine?
3. Has your dog become fussier or less interested in food recently?
4. Does your dog prefer wet or dry food?

Play

1. How important is play to your dog?
2. Who instigates play?
3. Has your dog's interest in play changed recently?
4. What does your dog do after a short play session?
5. What kind of play does your dog like?

Handling

1. Is your dog happy to wear a harness and collar? Has this changed?
2. Is your dog happy to be picked up? Has this changed?
3. Does your dog like to be cuddled?
4. Who instigates cuddling?
5. Is there anywhere your dog does not like to be touched?
6. Does your dog's tolerance for handling change throughout the day?
7. Is your dog happy to be groomed?

Social Behaviour

1. Does your dog like other dogs? Has this changed?
2. Does your dog like people? Has this changed?
3. Does your dog seem less social recently?
4. Does your dog take themselves away to a quiet place? How frequently?

5. Does your dog hide away when the house is busy?
6. Have any core relationships changed for your dog?

Other

1. Is your dog happy to jump on and off furniture? Has this changed?
2. Does your dog ever seem stiff or hesitant to get out of bed?

Index

Activity sharing, 65
Agoraphobia, 83
Animal ethics, 11
Animal Welfare Act (2006), 8
Animal Welfare Assessment Grid (AWAG), 8–9
Arousal, 40, 55–56
Association of Cats and Dogs Home (ACDH), 23

Balanced methods, 48
Behaviour, 4
    problems, 4
    case study, 114–115
    changes, 16–17, 21, 110
    and cognitive health, 115
    function identification, 52
    gastrointestinal issues, 115
    modification techniques, 11
    musculoskeletal pain, 113–114
    pain and, 112–113
Bloodhound, 25
Border Collies, 24, 25, 79–80
Brachycephalic Obstructive Airway Syndrome (BOAS), 109
Breeding, 109–110
Breed selection, 23
    behavioural implications, 24–25
    natural behaviours, 25–26
    physical considerations, 23–24
Breeds, classification of, 17
Breed-specific behaviours, 79–81
British Small Animal Veterinary Association (BSAVA), 108

Calmness, 67, 73, 75, 89
Canine Arthritis Management (CAM), 112
Canine cognitive dysfunction syndrome (CCDS), 115
Canine predatory sequence (CPS), 78–79
Car travel, 99–100
    CBD oil, 102
    comfortable access, 101
    desensitization, 102
    games, 101–102
    pheromone collars, 102
    reduction, 100
    safety signals, 100
CBD oil, 102
Chihuahuas, 25
Classical conditioning, 45–46

Cognitive dysfunction syndrome (CDS), 115
Confidence building, 40–41, 83–84
Consent-based handling, 38–39
Constructing Canine Consent (Jones), 1
Consultation, 117–118
Co-operative care, 119–124
Cost, 3–4
Counter-conditioning, 97–98
Crossbreed dog, 24

Dachshunds, 23–25
Dangerous Dogs Act (DDA), 36
Desensitization, 97–98, 102, 119–124
Digging, 81
Discomfort, 106, 107
Disobedient, 45
Distractions, 57–58
DNA sequencing, 14
Dog breeder selection, 22–23
Dog Breeding Licence, 22
Dog walkers, 93–94
Doggy day-care, 92–93
Dog-human attachment, in home
    authoritarian people, 62
    authoritative person, 62
    baby, 64
    children, 63–64
    emotional wellbeing
        activity sharing, 65
        attention, 64–65
        rules and boundaries, 65–66
        safe haven, 65
        sleep, 64
    permissive people, 62
    pet parenting, 62
    styles, 63
    visitors, 66
        calmness practising, 67
        environmental management, 66
        human behaviour, 67
        touching with consent, 67
        treat and retreat, 67–68
Dog-other animals attachment
    in home, 70
        with cats, 73
        inter-dog aggression, 70
        with other animals, 74
        with other dogs, 70–72

Dogs Trust National Dog Survey, 2
Domestication, 13
 behavioural changes, 16–17
 of dogs, 14
 genetics changes impact on, 15–16
 modern implications, 17–18

Electronic collars (e-collars), 54–55
Emotional valence, 55–56
Emotional wellbeing, 11, 12, 21
 equipment, 52–53
  electronic collars (e-collars), 54–55
  ethical considerations, 53
  physical considerations, 54
 learning environment, 45
  arousal, 55–56
  classical conditioning, 45–46
  distractions, 57–58
  emotional valence, 55–56
  operant conditioning, 47–48
  positive punishment methods, 49
  positive reinforcement methods, 49
  training methods, 48–50
  trigger stacking, 57
 leaving dog alone, 69–70
Empathy, 107
Engage/disengage exercise, 97–98
Environmental enrichment, 74–75
Equipment, 52–53
 electronic collars (e-collars), 54–55
 ethical considerations, 53
 physical considerations, 54

Fetch, 81–83
Free-ranging dogs, 18–19
French Bulldogs, 22, 23

Genetics changes, 14–15
 glucocorticoids (GCs), 15
 impact on domestication, 15–16
 oxytocin (OT), 15
German Shepherds, 25
Glucocorticoids (GCs), 15
Greetings, 88
 dog-to-dog greetings, 89
 people, 91
  appeasement signals, 92
  human body language, 92
  muzzle train, 91
 unfamiliar dogs, home, 89
 unfamiliar dogs outdoors, 89
  calmness, 89
  follow the leader, 90
  off-lead interaction, 90–91
  parallel walking, 90
  preparation, 89
  three-second greeting, 90

Handling
 at grooming salon, 118
  groomers guidance, 119
  high-welfare salon, 118–119
 low stress, 116–117
 in veterinary clinic, 116
Hunting, 80–81
Huskies, 25

Intensive breeding, 33
Inter-dog aggression, 71
International Association for the Study of Pain (IASP), 106

Koehler Method, 48, 49

Labrador Retrievers, 2, 25
Learning environment
 emotional wellbeing, 45
  arousal, 55–56
  classical conditioning, 45–46
  distractions, 57–58
  emotional valence, 55–56
  operant conditioning, 47–48
  positive punishment methods, 49
  positive reinforcement methods, 49
  training methods, 48–50
  trigger stacking, 57
Least inhibitive, functionally effective (LIFE) model, 44, 50–53
Least intrusive minimally aversive (LIMA) model, 50–52
Low stress handling, 116–117
Lucy's Law, 33–34

Maslow's hierarchy of needs, 9–11
Mud-wallowing, 81
Muzzle training, 84–88

Negative punishment, 48
Negative reinforcement, 47

Obedience, defined, 45
Obesity, 107–108
Off-lead interaction, 90–91
Operant conditioning, 47–48
Osteoarthritis, 108
Outdoor environment, 77
 exhibiting natural behaviours, 77–78
  canine predatory sequence (CPS), 78–79
  herding dogs, 79–80
  scent work and retrieving, 79
  sniffing, 78
  tracking, 80–81
Oxytocin (OT), 15, 16

Paedomorphosis, 17
Pain, 107
    and behaviour, 112–113
        case study, 114–115
        change, 110
        and cognitive health, 115
        gastrointestinal issues, 115
        musculoskeletal pain, 113–114
    chronic, 112
    defined, 106
    dermatological, 108–109
    economic pressures, 110
    reflections, 111
    veterinary dilemma, 110
    and wellbeing, 111–112
Pavlov's dogs, 46
People's Dispensary for Sick Animals (PDSA), 2
Pet parenting styles, 63
Pheromone collars, 102
Physical changes, 17
Physiological needs, 9–11
Positive punishment, 48, 49
Positive reinforcement, 48, 49
Positive restriction, 72
Predictable routine, 68–69
Pugs, 23
Puppy Contract, 22
Puppy farms, 33–36
Purebred dogs, 24

Quality-of-life (QOL) assessment, 112

Reactivity, 94
    caregivers mindset, 95
    counter-conditioning, 97–98
    desensitization, 97–98
    engagement exercises, 95–97
    walks, 98–99
Rehoming and abandonment crisis, 3, 23
Rescue, 23
Reset method, 72

Sadness, 107
Safe haven, 37–38, 65, 74
Secure attachment style, 39–40
Sensitivity, 16
Separate safe space, 71–72
Separation related problem (SRP), 70
Sheepballs, 80
Shih Tzus, 23
Skinner's Rats, 47
Sleep, 64
Slip leads, 58
Sniffing, 78
Social isolation, 22
Socialization, 16

Stranger situation test, 49
Surgical birth, 109–110
Swimming, 81

Terriers, 81
Trauma-informed care (TIC), 28, 29
    connections, 30
        kind and ethical training methods, 40
        secure attachment style, 39–40
    emotions management, 30–31, 40
        arousal, 40
        confidence building, 40–41
    safety, 29–30, 36
        consent-based handling, 38–39
        exposure to triggers management, 39
        needs assessment, 37
        resource, 38
        safe haven, 37–38
        support plan, 37
Traumatized dogs, 2, 23, 28–29, 42
    20-minute walk, 2
    behaviours, 28, 29
    characteristics, 29
    by criminal activity, 36
    overseas rescue dogs, 31
        behaviour and welfare concerns, 31–33
        puppy farms, 33–36
    trauma-informed care (TIC) (*see* Trauma-informed care (TIC))
Trigger stacking, 57

UK dog demographic, 2–3, 5

Vet's perspective
    breeding, 109–110
    conformation, 109–110
    discomfort, 107
    empathy, 107
    obesity, 107–108
    osteoarthritis, 108
    pain, 107
        behavioural change, 110
        dermatological, 108–109
        economic pressures, 110
        reflections, 111
        veterinary dilemma, 110
    sadness, 107
    surgical birth, 109–110
    veterinary practice, 107
Violations, 8

Walk refusal, 83
Welfare, 11–12
    Animal Welfare Assessment Grid (AWAG), 8–9

Welfare (*Continued*)
    definition, 7
    Five Welfare Needs, 8
    good indicators, 7–8
    Maslow's hierarchy of needs, 9–11
    poor indicators, 7
Wellbeing measurement, 6, 7
Working breeds, 24

www.ingramcontent.com/pod-product-compliance
Lightning Source LLC
Chambersburg PA
CBHW040541220526
45473CB00016B/2994